Nelson Advanced

Make the Grade

AS and A
Physics

Chris Honeywill

D1147769

Published in 2002 by:
Nelson Thornes Ltd
Delta Place
27 Bath Road
CHELTENHAM
GL53 7TH
United Kingdom

02 03 04 05 06 / 10 9 8 7 6 5 4 3 2 1

A catalogue record for this book is available from the British Library

ISBN 0 17 448280 9

Illustrations and typesetting by Mathematical Composition Setters Ltd, Salisbury, Wiltshire

Printed in Croatia by Zrinski.

The practice questions and mark schemes are based upon existing Edexcel Foundation questions.

Every effort has been made to trace all the copyright holders, but where this has not been possible the publisher will be pleased to make any necessary arrangements at the first opportunity.

Acknowledgements

The author would like to thank the staff at Edexcel, especially Frances Kirkman, Mary Gilbert and David Hartley for their help and advice. Particular thanks go to Ian Craig for his in-depth comments on the initial manuscript; to Mark Ellse for giving me the inspiration to write and Martyn Bassant for all his encouraging comments along the way.

Contents

Introduction

How to use this book

This book has been written to be used alongside the four Nelson Advanced Science Physics student books. It aims to help you develop your study skills, to make your learning more effective and to give you help with your revision. You may be entered for Unit Tests at different stages of your course, so you need to be prepared right from the beginning.

The chapters of this book are arranged in the order of the Units of the specification (or syllabus, as it used to be called). Chapters 1–3 deal with the AS content. Chapters 4–6 help you prepare for the A2 assessments if you are studying for Advanced GCE Physics.

The format of each section is:

- **Introduction** – gives an overview of the concepts of the material covered in that section
- **Things to understand** – the important points relating to the material covered
- **Things to learn** – the equations, laws, definitions and experiments that you need to learn
- **Checklist** – to help you check that you have covered and understood everything in that section
- **Testing your knowledge and understanding** – a multiple choice 'quick test', some worked examples and a number of practice assessment questions for you to try. Full answers to both the 'quick test' and the practice assessment questions are in the Answers section which follows Chapter 6.

Assessment of the physics specification

The Edexcel specification for AS physics is assessed by four written tests:

Test	Type and purpose	Duration
PHY1	About 8 structured questions on Unit 1	1 h 15 min
PHY2	About 8 structured questions on Unit 2	1 h 15 min
PHY3/01	1 structured question on your chosen topic	45 min
PHY3/02	2 practical questions	1 h 30 min

Note: All Tests may include assessment of your understanding of material in the 'General Requirements' section of the specifications (see Appendix 1).

Test PHY1 **only** examines material in Unit 1.
Test PHY2 mainly examines the content of Unit 2 but some questions might refer to physical principles contained in Unit 1.
Test PHY3/01 consists of four structured questions, one relating to each of the four topics in Unit 3. You only have to answer **one** of these.
Test PHY3/02 is based on the content of Units 1 and 2. It examines practical laboratory skills: planning; implementing; analysing evidence and drawing conclusions; evaluating evidence and procedures. One of the questions may involve drawing a graph and part of either question may require you to use your experience of practical datalogging techniques.

Note: All these Tests may also include assessment of your understanding of material in the 'General Requirements' section of the specifications (see Appendix 1).

In addition to the above tests, the Edexcel specification for Advanced GCE physics is assessed by four more written tests:

Test	Type and purpose	Duration
PHY4	About 8 structured questions on Unit 4	1 h 20 min
PHY5/01 PHY5/02	About 6 structured questions on Unit 5 3 practical questions	1 h 1 h 30 min
PHY6	4 synoptic questions	2 h

Test PHY4 mainly examines the content of Unit 4. It assumes that Units 1 and 2 have been studied but does not examine their content again in detail.

Test PHY5/01 assumes that Units 1, 2 and 4 have been studied. Much of the content of Unit 5 builds on these previous units and this is reflected in the questions, although all are set in the context of Unit 5.

Test PIIY5/02 is based on material from Units 1, 2, 4 and 5 of the specification and is designed to build on the practical laboratory skills already tested in Test PHY3/02. At least one of the questions involves drawing a graph, which may involve the use of logarithms, and part of any question may require you to use your experience of practical datalogging techniques.

Test PHY6 examines your accumulated understanding of the whole Advanced GCE specification. This test is answered in a separate answer book. Question 1 involves the analysis of a passage adapted from a scientific or technological book or journal. Question 2 tests your understanding and applications of the principles drawn together in Unit 6. Questions 3 and 4 examine material from the rest of the specification (Units 1, 2, 4 and 5) and each of these questions will require an understanding of principles from more than one Unit.

Data, formulae and relationships

A selection of data, formulae and relationships will be printed at the end of each test paper. Appendix 2 gives the full list.

 Study skills

Revision is a personal activity. What works best for you may not be so effective for someone else. However there are some golden rules.

1 Revise little and often.

2 Revise actively – do not sit and stare at your notes or this book. Write down important points or use a highlighter to mark important passages in your notes or in this book (but only if you own it!).

3 Work out answers to the questions and then check them with those given.

4 Help each other. Explaining a point of physics to another student is a good way of clarifying your own understanding. Test each other by asking simple questions, such as formulae, definitions, units and experimental descriptions.

Do not leave your revision until the last minute. Revision should take place throughout the whole course.

Here are some suggestions to help you study and prepare for your Unit Test papers:

Daily tasks

After each lesson check that your notes are complete. Try spending 10 to 15 minutes looking through them. If there is something that you do not understand:

● Read the relevant part in this book or your textbook and, if necessary, add to your notes so that they will be clear when you read them again.

● Discuss the problem with another student.

● If you still have difficulty, ask your teacher as soon as you can.

The more you contribute to solving each problem, the deeper and longer lasting your understanding of it will be.

Weekly tasks

● Look through your notes. Highlight important parts.

● Read through the relevant parts of this book and make notes and/or highlight important points.

● Complete any homework assignments.

End of section tasks

When your teacher has completed a section of work, you should revise that material thoroughly. To do this:

● Work through your notes alongside a copy of that part of the Edexcel specification (syllabus).

● Summarise your notes to the bare essentials.

● Work through the relevant material in this book. Discuss any difficulties with other students.

● Attempt **all** the 'quick test' questions for that section of work.

The specification can be found on Edexcel's web site at www.edexcel.org.uk

 Preparing for the Unit Tests (examinations)

If you have followed the previous advice, you will find it easier to prepare for the assessment tests. Bear in mind that Unit Tests 1 and 4 take place on the same day, as do Unit Tests 2 and 5. You will also be taking tests in other subjects, so you should aim to start your final revision at least four weeks in advance.

● Try spending about 30 minutes revising one subject. Then switch from physics to another subject.

● Take regular breaks.

● Revise actively with pen, highlighter and paper.

When you have fully revised the material in a Unit, read through the 'worked examples' and attempt the 'practice assessment questions' provided in this book. Mark your work using the answers in the Answers section which follows Chapter 6 – or better still mark a friend's work and let him or her mark yours. Then:

● Work out where you went wrong.

● If you obtained low marks for a particular section, go back to your notes and textbooks and look over that section before having another attempt.

Spreading revision this way over the whole course will reduce stress and will guarantee a better grade than you would obtain by leaving it all to a mad dash at the end. Physics is a subject in which knowledge is built up gradually. The more thoroughly you work in the earlier stages, the easier and more enjoyable you will find the study of physics.

The day of the Unit Test

If you have followed the advice given here, you should feel confident that you will be able to do your best. Some people find it helpful to spend a little time looking over some physics before going into the test, others prefer to keep their minds clear for the task ahead.

Check that you have:

> Don't take a red pen with you as the awarding body doesn't allow you to use this colour – the examiners use red for marking the papers.

- Two or more blue or black pens and several pencils.
- Your calculator – if the batteries are old replace them beforehand.
- A watch – try putting it on the desk in front of you.
- A ruler.
- A good luck charm, if it helps.

 Tackling the question paper

- Work steadily through the paper starting at question 1.
- The questions in the seven Tests associated with Units 1–5 are answered in the spaces provided on the question paper itself. If you need more room for your answer, look for space at the bottom of the page, at the end of the question or after the last question.

> If you use space at the bottom of the page let the examiner know by adding 'continued below' or 'continued on page ●●●' if space is used elsewhere.

- Use the amount of space given for each answer as a guide to how much you should write. If a question has three lines for the answer, do not write an essay. Work out the essential points that need to be made, and check them against the number of marks to be awarded.
- Do not repeat the question in your answer.
- Pace yourself so that you neither run out of time nor have masses of time to spare at the end. If you get stuck, do not waste time. Make a note of the question number and part that caused you difficulty and go on. Later, if you have time, go back and try that part again.
- Using correcting fluids can waste time while you wait for it to dry. Frequently an examiner sees a thick crust of white with nothing written on it, and wonders whether some marks might have been given for what had originally been written. Rather than using correcting fluid, neatly cross out what you have written. If, later, you realise that what you had first written was correct, write 'ignore crossing out' beside the work that you had crossed out. The examiner will then mark it.

Terms used in the Tests

It is important that you understand what the examiners want.

Some of the terms that are often used in questions are explained below:

- **Calculate**: a numerical answer is obviously required! Show your working and set your work out clearly. Don't forget the units.
- **Comment**: make sure what you write is relevant. Judge amount of detail required from marks/space.
- **Complete**: add to (circuit) diagrams and/or tables.

- **Define**: you can define quantities by their equations but remember to explain any symbols used.
- **Describe**: give the main points as precisely as possible. Labelled diagrams can help and are essential when describing experiments.
- **Explain**: give some reasoning or refer to theory. A labelled diagram will often improve your answer. Judge amount of detail required from marks/space.
- **Plot**: use scales on graph paper and be precise. Show data points either as a cross or a dot surrounded by a small circle.
- **Show that**: show all your working and give your answer to one more significant figure than the approximate value stated in the question. It is very likely that the stated value will be needed in a calculation later in the same question. So even if you can't do this part, you can still attempt the next!
- **Sketch**: use labelled axes but only add axes values if told to do so. Sketch roughly but carefully.
- **State**: a brief sentence giving the required facts. No explanation is required.
- **Suggest**: there is often no single correct answer. Credit is given for good physics reasoning.
- **Use the graph**: usually this involves finding either the gradient or the area. Remember that both of these quantities are likely to have units.

and finally

Examiners **do** try wherever possible to give you marks rather than looking for ways to take them away.

Be prepared, be confident and you will do your best, which is all that anyone can ask of you.

1 Mechanics and radioactivity

Part ❶ Mechanics

 Introduction

Mechanics describes the effects of forces when they act on bodies that are either at rest (statics) or in motion (dynamics). As you study mechanics, you become aware of the conditions needed for a body to be in equilibrium and how a body moves when acted upon by a resultant force. You use both graphs and equations to describe such motion. You learn about the behaviour of colliding bodies and understand the energy exchanges that are taking place. Most importantly, you find the true meaning of work (at least as it applies to the world of the physicist!).

 Things to understand

Density

the density of a material is its mass for a unit volume, usually 1 m^3

- the densities of solids are usually larger than those of liquids
- the densities of liquids are much larger than those of gases

Motion in a straight line

- distance (a scalar) measured along a straight line in a particular direction is called displacement (a vector)
- speed (a scalar) is the distance moved per second whereas velocity (a vector) is displacement per second
- velocity is speed in a given direction
- the directions must be taken into account when vectors are added or subtracted (Figure 1.1)
- acceleration (a vector) is the rate of change of velocity or change in velocity per second
- displacement–time, velocity–time and acceleration–time graphs are a useful method for displaying information about the motion of a body (Figure 1.2)
- the gradient at any point on a displacement–time graph is the velocity at that point sometimes called the instantaneous velocity (Figure 1.2)

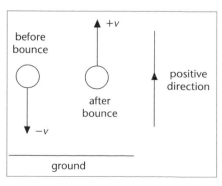

Fig 1.1 *This ball's velocity changes from −v to +v as it bounces so its velocity changes by [+v − (−v)] = 2v*

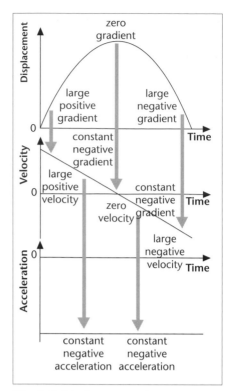

Fig 1.2 *Displacement–time, velocity–time and acceleration–time graphs for a ball thrown vertically upwards (upwards taken as positive)*

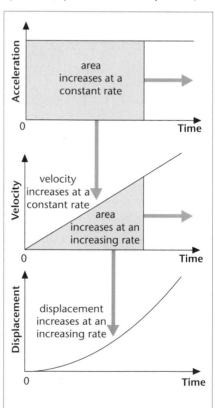

Fig 1.3 *Acceleration–time, velocity–time and displacement–time graphs for a ball dropped from rest (downwards taken as positive)*

- the gradient at any point on a velocity–time graph is the acceleration at that point (Figure 1.2)
- the area enclosed by an acceleration–time graph is the change in velocity (Figure 1.3)
- the area enclosed by a velocity–time graph is the change in displacement (Figure 1.3)
- the following equations describe the motion of an object moving with constant acceleration in a straight line (Table 1.1 defines the symbols used and gives their units)

$$x = \tfrac{1}{2}(u + v)t$$
$$v = u + at$$
$$x = ut + \tfrac{1}{2}at^2$$
$$v^2 = u^2 + 2ax$$

v	final velocity	m s^{-1}
u	initial velocity	m s^{-1}
x	displacement	m
a	acceleration	m s^{-2}
t	time	s

Table 1.1 *Symbols used in the equations of motion*

Projectile motion

- when air resistance is removed, all objects fall with the same acceleration
- the acceleration of a projected object is vertically down and equal to the acceleration of free fall *g* throughout its flight, whether the object is on its way up, at the top of its path or on its way down
- an object that is projected horizontally falls to the ground with the same acceleration (*g*) as one falling vertically; horizontal and vertical motions of an object are independent of each other
- the curved path, called a parabola, of a horizontally projected object is the result of a constant horizontal velocity (when air resistance is zero) combined with a uniform vertical acceleration

Forces

- force (a vector) involves the push or the pull of one thing on another
- forces can be gravitational, electrostatic, electromagnetic or nuclear
- both tension (Figure 1.4) and weight (the gravitational pull of the Earth on an object) are forces
- the centre of gravity of a body is the point where all its weight appears to act
- a sketch of a single object that shows all the forces acting on it is called a free-body force diagram
- in situations where the forces do not have the same line of action, a vector diagram can be used to find their resultant; the single force that could replace them all and have the same effect
- in some situations, when analysing the forces acting, it is helpful to split up a single force into two perpendicular components (Figure 1.5)

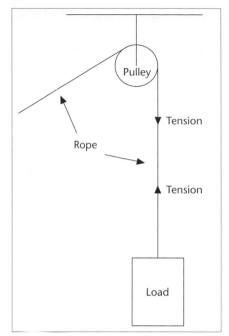

Fig 1.4 *The tension in the rope acts upwards on the load and downwards on the pulley*

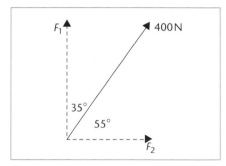

Fig 1.5 *Component*
$F_1 = 400\,N \times \cos 35° = 330\,N$;
Component
$F_2 = 400\,N \times \cos 55° = 230\,N$

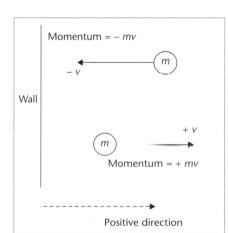

Fig 1.8 *The ball's momentum changes from −mv to +mv as it hits the wall so its momentum changes by* $[+mv − (−mv)] = 2\,mv$

- forces always occur in pairs; when body A exerts a force (an action) on body B, body B automatically exerts a force (a reaction) on body A
- action and reaction forces   are equal in magnitude
                                    are opposite in direction
                                    act on different bodies
                                    are of the same type (e.g. both gravitational)
                                    act for the same length of time
                                    have the same line of action
- action and reaction forces cannot cancel each other as they act on different bodies

Forces and moments

- a force can have a rotational effect on a body; the moment of a force is a measure of its rotational effect

      moment of F about O = F × perpendicular distance from F to O

  moments can be clockwise or anticlockwise (Figure 1.6)
- a couple (Figure 1.7) consists of two equal and opposite non-aligned forces

      moment = one of the forces F × perpendicular separation d

- for a body in equilibrium

  the sum of the forces in any direction must be zero

  the sum of the moments about any point must be zero (i.e. Σ clockwise = Σ anticlockwise)

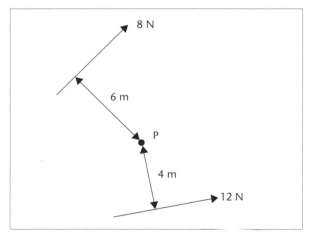

Fig 1.6 *The moments of these forces about the point P are both 48 N m*

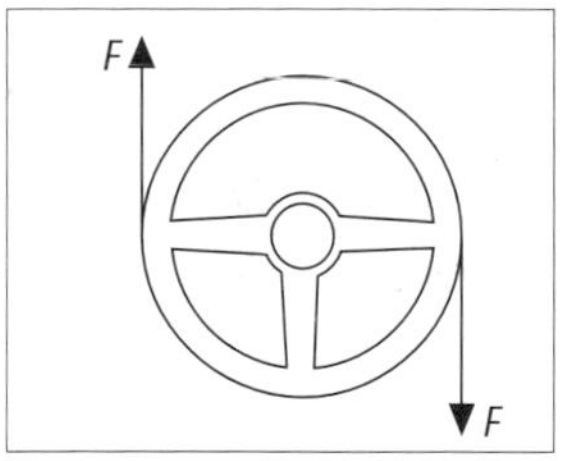

Fig 1.7 *Two equal and opposite forces exerting a couple on a steering wheel*

Forces and motion

- the resultant force on either a stationary body or a body moving with constant velocity is zero
- a resultant force is needed to accelerate a body; for the same force, a large mass will accelerate less than a small one
- the acceleration of a body is proportional to the resultant force and occurs in the same direction as this force
- the momentum of a body is the product of its mass and its velocity; it is a vector quantity in the same direction as the velocity (Figure 1.8)

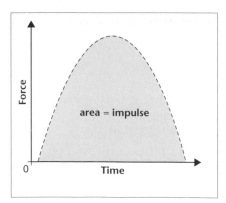

Fig 1.9 *The area under a force–time graph is the impulse of that force*

- the rate of change of momentum of a body equals the resultant force acting on it and occurs in the same direction as this force

$$(mv - mu)/t = F = ma$$

- the impulse of a force is the product of the force and the time for which it acts and, particularly for a changing force, can be found from the area under a force–time graph (Figure 1.9)

- the impulse of a force equals the change in momentum that it produces

- when two bodies collide, they exert equal and opposite impulses on each other (equal and opposite forces for the same length of time) – hence, they undergo equal and opposite changes in their momentum so there is no change in the total momentum and so momentum is conserved

Mechanical energy

- work (a scalar) is done by a force when it causes motion

  work done = average force × distance moved in the direction of the force

  = applied force × displacement parallel to force

- work done is equal to the area under a force–displacement graph even when the force varies

- energy is transferred when work is done; the system doing work loses energy whereas the system having work done on it gains this energy

- the total amount of energy in an isolated system remains constant

- a moving body possesses kinetic energy due to its motion

  kinetic energy = $\frac{1}{2}mv^2$

- kinetic energy is conserved in an elastic collision, whereas some kinetic energy is transferred to other forms in an inelastic collision

- a body raised above the Earth's surface possesses gravitational potential energy due to its position: for a mass raised through a distance Δh

  change in gravitational potential energy = $mg\Delta h$

- the efficiency of a system indicates the proportion of the energy input that can be usefully used

  efficiency = useful output/total input

- power is the rate at which energy is transferred *or* the rate at which work is being done

- when an applied force is causing motion, the power developed by the force is

  power = applied force × velocity

 Things to learn

You should learn the following for your Unit PHY1 Test. Remember that it may also test your understanding of the 'general requirements' (see Appendix 1).

Equations that will *not* be given to you in the test

☐ density = mass/volume

$\rho = m/V$ ρ = density

☐ (average) velocity = (total) displacement/time taken

$\frac{1}{2}(u + v) = x/t$ u = initial velocity v = final velocity

☐ acceleration = change in velocity/time taken

$a = (v - u)/t$

☐ resultant force = mass × acceleration

$F = ma$

☐ momentum = mass × velocity

$p = mv$ p = momentum

☐ work done = applied force × distance moved in the direction of the force

$\Delta W = F\Delta x$ Δx = change in displacement

☐ power = energy transferred/time taken = work done/time taken

$P = W/t$

☐ weight = mass × gravitational field strength

weight = mg

☐ kinetic energy = $\frac{1}{2}$ × mass × speed^2

k.e. = $\frac{1}{2}mv^2$

☐ change in gravitational potential energy = mass × gravitational field strength × change in height

Δp.e. = $mg\Delta h$ where Δ = 'change in'

Laws

☐ Newton's first law: a body will remain at rest or continue to move with a constant velocity as long as the forces on it are balanced

☐ Newton's second law: the rate of change of momentum of a body is directly proportional to the resultant force acting on it and takes place in the same direction as the resultant force

☐ Newton's third law: while body A exerts a force on body B, body B exerts an equal and opposite force on body A

☐ principle of moments: if a body is in equilibrium, the sum of the moments about any point must be zero

☐ conservation of momentum: provided no external forces act, the total momentum of a system of objects remains constant

☐ conservation of energy: the energy content of a closed or isolated system remains constant

General definitions

☐ force: a push or a pull involving at least two bodies; something that can cause a body to accelerate

☐ resultant force: the single force that could replace all forces acting and have the same effect

☐ one newton: the resultant force that gives a mass of 1 kg an acceleration of 1 m s^{-2}

- moment of a force about a point: the product of the force and its perpendicular distance from that point
- centre of gravity: the point where all the weight of the body appears to act
- impulse: the product of a force and the time for which it acts
- power: the rate of doing work
- elastic collision: a collision in which kinetic energy is conserved
- inelastic collision: a collision in which kinetic energy is not conserved (e.g. some may be dissipated as thermal energy)

Word equation definitions

Use the following word equations when asked to define:

- density = mass/volume
- (average) speed = (total) distance travelled/time taken
- (average) velocity = (total) displacement/time taken
- acceleration = change in velocity/time taken
- momentum = mass × velocity
- work done = force × distance moved in the direction of the force
- efficiency = useful energy output/total energy input

Experiments

Several mechanics experiments involve the measurement of velocity or acceleration. In many cases, suitable measurements can be taken in a number of ways using a variety of apparatus. Whatever method you describe, you must explain what is being measured and how these measurements are then used. You will not get any marks for 'the velocity was obtained by passing the object through an intelligent timer'.

The first two experiments describe different methods for measuring velocity and acceleration – only learn one of these methods, the one with which you are most familiar.

1. Measuring velocity

In all methods, a measured distance is divided by a measured time.

Method 1 Using an electronic timer operated by a light gate
Attach a card of measured length centrally to the top of the vehicle. Arrange for the card to block a light gate's beam as it passes through it (Figure 1.10).
Electronic timer measures how long card takes to pass through beam.
Calculate vehicle's average velocity as it passes the light gate, *v* = length of card/interruption time.

Method 2 Using tickertape
Attach a length of tickertape to the back of the vehicle which pulls it through a ticker-timer machine.
Measure the length of 10 adjacent gaps between the dots with a metre rule.
Time taken = 0.2 s (tickertimer makes 50 dots each second).
Calculate vehicle's average velocity during this time using *v* = length of 10 gaps/(0.2 s).

Method 3 Using a video camera
Video the vehicle moving along in front of a calibrated scale.

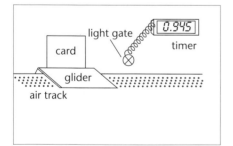

Fig 1.10 *The timer records the time taken for the card to pass through the light gate's beam*

Play the video back a frame at a time.
Measure how far the vehicle advances between frames from the scale.
Time between frames is 0.04 s (video camera takes 25 frames each second).
Calculate vehicle's average velocity between frames using v = distance moved between frames/(0.04 s).

☐ 2. Measuring acceleration

In all methods, at least two velocities are found and the change in velocity is divided by the measured time for this change to occur.

Method 1 Using an electronic timer operated by two light gates
Attach a card of measured length centrally to the top of the vehicle.
Arrange for the card to block the beams of two light gates as it passes through them.
Timer measures how long card takes to pass through each light beam (t_1, t_2).
Record time for vehicle to pass between the two gates using a stopwatch (t_3).
Velocity difference = length of card/t_2 – length of card/t_1
Acceleration = velocity difference/t_3

Method 2 Using a timer, a light gate and a double interrupter card
Attach two cards of the same measured length symmetrically to the vehicle.
Arrange for the cards to block the light gate's beam as they pass through it (Figure 1.11).
Timer measures how long each card takes to pass through the light beam (t_1, t_2).
Timer also measures time interval between the start of the two interruptions (t_3).
Velocity difference = length of card/t_2 – length of card/t_1
Acceleration = velocity difference/t_3

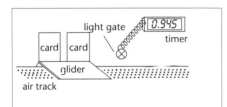

Fig 1.11 *The time taken for each card to pass through the light gate's beam is recorded by the timer, together with the time interval between the start of the two interruptions*

Method 3 Using tickertape
Attach a length of tickertape to the back of the vehicle which pulls it through a ticker-timer machine.
Measure the length of the first five adjacent gaps with a metre rule.
Time taken = 0.1 s (tickertimer makes 50 dots each second).
Calculate vehicle's average velocity during this time using v = length of 5 gaps/(0.1 s).
Repeat for several consecutive sets of five adjacent gaps.
Plot a graph of velocity against time.
Acceleration = gradient of graph.

Method 4 Using a video camera
Video the vehicle moving along in front of a calibrated scale.
Play the video back a frame at a time.
Measure how far the vehicle advances between frames from the scale.
Time between frames is 0.04 s (video camera takes 25 frames each second).
Calculate vehicle's average velocity between frames using v = distance moved between frames/(0.04 s).
Repeat for several consecutive frames.
Plot a graph of velocity against time.
Acceleration = gradient of graph.

☐ 3. Measuring the acceleration of free fall

The acceleration of free fall can be measured by dropping a double interrupter card through a light gate.

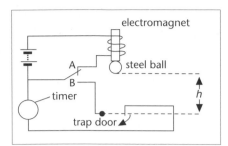

Fig 1.12 *With the switch in position A, the electromagnet attracts the ball bearing*

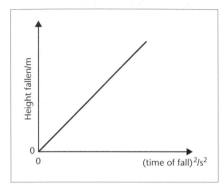

Fig 1.13 *Acceleration of gravity = 2 × gradient*

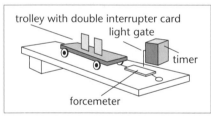

Fig 1.14 *Finding the acceleration a produced by the force F*

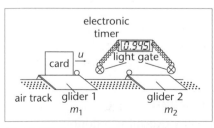

Fig 1.15 *Investigating a collision where two gliders join together*

The following method involves timing an object as it falls from rest ($u = 0$) over a measured distance and using the equation $x = \frac{1}{2}at^2$ to find its acceleration. A best-fit straight-line graphical method is used to average the results and the acceleration is found from the gradient of the graph. The electromagnet attracts the ball bearing (Figure 1.12).

Measure the height h from the bottom of the ball to the trapdoor switch. When the switch is moved to B, the ball is released and the timer starts. The timer stops when ball hits and opens the trapdoor switch. Record the time of fall.
Find average time of fall t from at least three attempts.
Repeat for a range of different heights; tabulate values for h and t.
Plot a graph of h against t^2 to get a straight line through origin (Figure 1.13).
Comparing $h = \frac{1}{2}at^2$ with $y = mx + c$ shows that the gradient is $\frac{1}{2}a$.
So acceleration of free fall = 2 × gradient.

☐ 4. The relationship between force and acceleration for a fixed mass

This experiment involves applying different known forces to a fixed mass and measuring the acceleration that is produced.
The following method uses a forcemeter and a double interrupter card with light gate to measure these two quantities.

Tilt the runway so that, after an initial push, the trolley runs down it at a constant speed (no acceleration). The runway is now friction compensated.
Attach two cards of the same measured length symmetrically to the trolley so that they block a light gate's beam as they pass through it.
Use a forcemeter to apply a constant force F to the trolley (Figure 1.14).
The timer measures how long each card takes to pass through the light beam (t_1, t_2) and the time interval between the start of the two interruptions (t_3).
Acceleration a = (length of card/t_2 – length of card/t_1)/t_3
Repeat for a range of forces.
either
Plot a graph of acceleration against force.
A straight line through the origin shows that acceleration and force are directly proportional.
or
Calculate F/a for each force used.
If answers are the same, acceleration and force are directly proportional.

☐ 5. Conservation of linear momentum

This experiment involves measuring the velocities of two colliding bodies both before and after a collision.

The simplest collision to describe is where one body is initially at rest and the two bodies join together during the collision. The following method uses light gates to measure the velocities of two gliders colliding on an air track.

Attach a card of measured length centrally to the top of the glider on the left so that it blocks the beams of each of the light gates as it passes through them (Figure 1.15).
Start the left glider moving to the right so that it collides with and sticks to the other glider which is at rest between the two light gates.
Timer measures how long the card takes to pass through each of the two light gates (t_1, t_2).

Calculate the left glider's velocity u at the first light beam (u = length of card/t_1) and their combined velocity v at the second light beam (v = length of card/t_2).

Measure the mass of each glider (include the card) m_1, m_2.

Compare the momentum of the left glider before the collision $m_1 u$ with that of the joined gliders after the collision $(m_1 + m_2)v$.

Repeat with different initial velocity u and using gliders of different mass.

In all cases, momentum is conserved if $m_1 u = (m_1 + m_2)v$.

☐ 6. Elastic and inelastic collisions

The apparatus and method for this experiment is similar to that for experiment 5. A second set of results is obtained with spring buffers on the gliders so that they gently bounce off each other when they collide.

For all sets of results, calculate the kinetic energy before the collision ($\frac{1}{2}m_1 u_1^2$) and the total afterwards ($\frac{1}{2}m_1 v_1^2 + \frac{1}{2}m_2 v_2^2$).

Find the percentage of the initial kinetic energy remaining after the collision.

Compare these percentages for the two types of collision.

The collision with the spring buffers should be closest to 100% indicating that this collision is the closest to being elastic.

☐ 7. Efficiency of energy transfer

The following experiments measure how much of the initial stored energy is converted into kinetic energy.

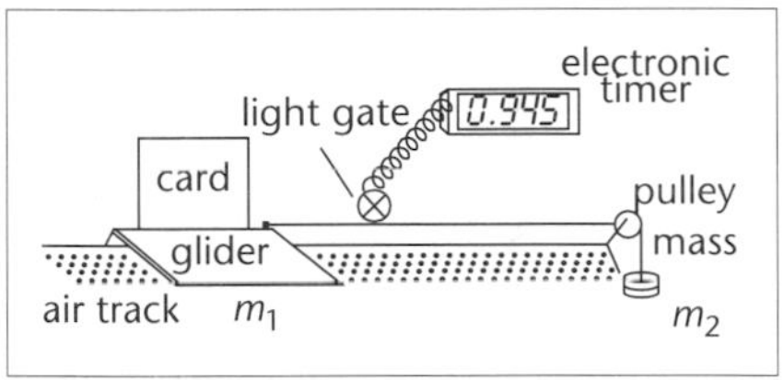

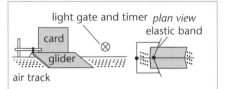

Fig 1.16 *What percentage of the gravitational potential energy becomes kinetic energy?*

Method 1 Gravitational potential energy to kinetic energy

Measure height h of the mass m_2 above the floor (Figure 1.16).

Position the light gate slightly further than h ahead of the glider.

Attach a card of measured length centrally to the top of the glider so that it blocks the light gate's beam as it passes through it.

Release the mass so it falls to the floor and accelerates the glider.

Glider moves at a constant velocity once the mass has hit the floor.

Timer measures how long the card takes to pass through the light gate.

Calculate the constant velocity v of the glider using length of card/interruption time.

Measure the mass of the glider (include the card) m_1 and the falling mass m_2.

Find the percentage of the gravitational potential energy ($m_2 g h$) of the falling mass that becomes kinetic energy [$\frac{1}{2}(m_1 + m_2)v^2$].

Repeat using different masses and release heights.

Method 2 Elastic potential energy to kinetic energy

Use a rule to measure the extension of the rubber band when stretched as shown in Figure 1.17 by different forces applied with a forcemeter.

Plot a force–extension graph for the rubber band.

Find the elastic potential energy stored in the elastic band for a number of extensions using the area under the graph.

Attach a card of measured length centrally to the top of the glider so that it blocks the light gate's beam as it passes through it.

Use the first extension to catapult the glider along the air track.

Timer measures how long the card takes to pass through the light gate.

Calculate the velocity v of the glider using v = length of card/interruption time.

Measure the mass of the glider (include the card) m.

Find the percentage of the elastic potential energy stored in the rubber band at this extension that becomes kinetic energy [$\frac{1}{2}mv^2$].

Repeat for the other extensions.

Fig 1.17 *What percentage of the elastic potential energy becomes kinetic energy?*

Checklist

Before attempting the following questions on mechanics, check that you:

❏ know the definition of density and can describe how to measure the densities of solids (including those with irregular shapes), liquids and gases

❏ know the meanings of the terms: distance, displacement, speed, velocity and acceleration

❏ can sketch displacement–time graphs for a body moving with a constant speed and for a body moving with a constant acceleration

❏ can sketch a velocity–time graph for a body moving with a constant acceleration

❏ know that the gradient of a displacement–time graph gives velocity and that of a velocity–time graph gives acceleration

❏ know that the area under a velocity–time graph gives the change in displacement

❏ can confidently use the equations of motion

❏ have learnt a description of an experiment to determine the acceleration of a freely falling object

❏ know that the parabolic path of a projectile results from a constant horizontal speed and a uniform vertical acceleration

❏ know that a force is a vector that acts at a particular point and that the resultant of a number of forces can be found using a vector diagram drawn to scale

❏ know how to draw free-body force diagrams and appreciate that the weight of a body is a force that acts through its centre of gravity

❏ know what is meant by the moment of a force and can state and apply the principle of moments

❏ know the conditions required for a rigid body to be in equilibrium and can use these to solve static force problems

❏ can calculate the momentum of a moving body

❏ know what is meant by the impulse of a force and can relate this to the change in momentum that it produces

❏ have learnt a statement of each of Newton's three laws of motion

❏ have learnt a description of an experiment to investigate the relationship between force and acceleration for a fixed mass

❏ know how Newton's second law of motion leads to the definition of the newton as the unit of force

❏ can identify pairs of action and reaction forces and know their properties

❏ have learnt a statement of the principle of conservation of momentum and appreciate that this principle follows on directly from a combination of Newton's second and third laws

❏ have learnt a description of an experiment to test the principle of conservation of momentum

❏ know the meanings of the terms: work, power, kinetic energy, gravitational potential energy and elastic potential energy

❏ have learnt a statement of the law of conservation of energy

❏ know the similarities and the differences between an elastic and an inelastic collision

☐ know how to find the efficiency of an energy transfer process

☐ are familiar with the 'general requirements' (see Appendix 1) and how they apply to the topic of mechanics

Testing your knowledge and understanding

Answers to these questions, together with explanations, are in the Answers section which follows Chapter 6.

Quick test

Select the correct answer to each of the following questions from the four answers supplied. In each case only one of the four answers is correct. Allow about 40 minutes for the 20 questions.

1 Which one of the following is NOT a base unit in the SI system?

    A Ampere   **B** Metre   **C** Newton   **D** Mole

2 In the equation $c = \sqrt{(k/\rho)}$, c represents a speed and ρ a density. The units in which the quantity k is measured are

    A kg m s^{-2} **B** $\text{kg}^{\frac{1}{2}}\text{ s}$ **C** kg m s^{-1} **D** $\text{kg m}^{-1}\text{ s}^{-2}$

3 The density of sand is 2500 kg m^{-3}. What is the volume in m^3 of 50 kg of sand?

    A 0.005   **B** 0.010   **C** 0.020   **D** 0.040

4 A body falls freely under gravity after being released from rest. Neglecting air resistance, which of the graphs in Figure 1.18 represents the variation of the height h of the body with time t.

5 The graph in Figure 1.19 shows how a trolley moved at a constant speed along a corridor.

Its constant speed was

    A 0.5 m s^{-1} **B** 2.0 m s^{-1} **C** 0.5 m s^{-2} **D** 2.0 m s^{-2}

6 A lunar landing module is descending to the Moon's surface at a steady velocity of 10 m s^{-1}. At a height of 120 m, a small object falls from its landing gear. If the Moon's gravitational acceleration is 1.6 m s^{-2}, at what speed in m s^{-1} does the object strike the surface of the Moon?

    A 10.0   **B** 19.6   **C** 22.0   **D** 202

7 A ball is suspended from an electromagnet attached to a trolley that is travelling as shown in Figure 1.20 at a steady speed of 1 m s^{-1}. The trolley is illuminated by a stroboscope that flashes at a regular rate. The ball is released and a stroboscopic photograph taken using a camera that is also moving to the right at 1 m s^{-1}.

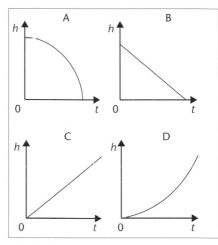

Fig 1.18

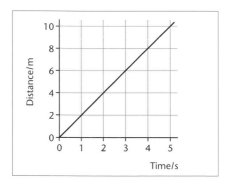

Fig 1.19

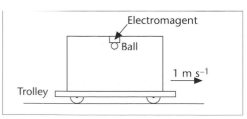

Fig 1.20

The photograph obtained is

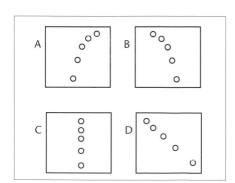

Fig 1.21

8 A rocket is accelerating upwards through the atmosphere due to the thrust of its jets. The force required by Newton's third law to pair with the weight of the rocket is the

    A Earth's gravitational pull on the rocket
    B thrust of the jets on the air
    C thrust of the air on the jets
    D rocket's gravitational pull on the Earth

9 Figure 1.22 shows a person sitting on a box that rests on the ground, together with a free-body force diagram for the box.
Which of the following statements is correct?

    A force Q > force P + force R
    B P is the push of the box on the Earth
    C Q is the pull of the box on the Earth
    D R is the push of the person on the box

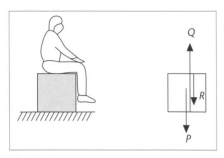

Fig 1.22

10 Figure 1.23 shows four systems, each having three coplanar forces acting at a point. The lengths of the force vectors represent the magnitudes of the forces. Which system of forces could be in static equilibrium?

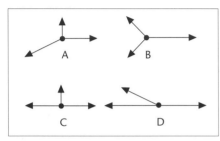

Fig 1.23

11 Figure 1.24 shows a rigid light rod PQ in equilibrium under the action of three forces F_1, F_2 and F_3. Which one of the following statements is true?

    A $F_1 \times d_1 = F_2 \times d_2$
    B $F_2 \times d_1 = F_3 \times d_2$
    C $F_1 \times d_1 = F_2 \times (d_1 + d_2)$
    D $F_2 \times d_1 = F_3 \times (d_1 + d_2)$

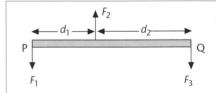

Fig 1.24

12 A uniform metal rod is 60 cm long and carries a weight of 10 N at one end and a weight of 20 N at the other end. The rod is supported in equilibrium by a knife-edge placed under it at a distance of 35 cm from the end carrying the 10 N weight as shown in Figure 1.25.

What is the weight of the rod?

    A 10 N   **B** 20 N   **C** 30 N   **D** 50 N

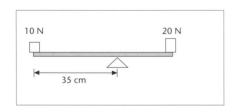

Fig 1.25

13 Figure 1.26 shows the forces acting on an aircraft as it climbs at a steady speed at an angle θ to the horizontal.

Which one of the following is true?

    A lift $\times \sin \theta$ = weight
    B the resultant force is zero
    C lift $\times \cos \theta$ = weight
    D the resultant force acts in the direction of the thrust

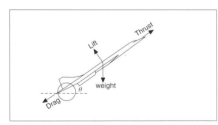

Fig 1.26

14 A body of mass 4 kg is accelerated from rest by a steady force of 5 N. What is its speed in m s^{-1} when it has travelled for 8 s?

    A 2.5   **B** 6.4   **C** 10   **D** 160

15 The graph in Figure 1.27 shows how a physical quantity Y, relating to a body of fixed mass, varies with time t.

The impulse experienced by the body over the time interval $(t_2 - t_1)$ is equal to the shaded area if Y is the

    A displacement of the body
    B resultant force acting on the body
    C momentum of the body
    D velocity of the body

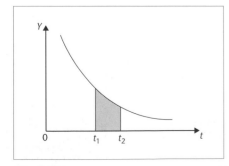

Fig 1.27

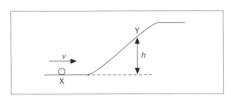

Fig 1.28

16 In laboratory experiments with colliding trolleys, the total momentum measured before a collision is hardly ever exactly the same as that measured after. This is because the total momentum is

    A altered by the action of external forces
    B conserved only when averaged over a large number of collisions
    C conserved only in perfectly elastic collisions
    D conserved only in totally inelastic collisions

17 An object of mass m passes a point X with a velocity v and rises up a frictionless incline to point Y that is at a height h above X as shown in Figure 1.28.

A second object of mass $\frac{1}{2}m$ passes X with a velocity of $\frac{1}{2}v$. To what height will it rise?

    A $h/8$ **B** $h/4$ **C** $h/2$ **D** h

18 A cricket ball of mass 150 g is caught by a fielder who stops the ball in a distance of 0.4 m with an average force of 27 N. What was the speed in m s^{-1} of the ball just before the fielder caught it?

    A 12   **B** 24   **C** 36   **D** 72

19 Two spheres of equal mass m travel towards each other with equal speed v on a smooth horizontal surface. They have a perfectly elastic head-on collision.

    A Before impact the total momentum of the two spheres is mv
    B After impact the total momentum of the two spheres is $2mv$
    C Before impact the total kinetic energy of the two spheres is $2mv^2$
    D After impact the total kinetic energy of the two spheres is mv^2

20 A girl of mass 50 kg runs up a flight of stairs 5 m high in 4 s. Taking the acceleration of gravity as 10 m s^{-2}, what is her power rating in W in raising herself through this vertical height?

    A 40.0   **B** 62.5   **C** 400   **D** 625

Worked example

Study the following worked examples on mechanics carefully. Make sure you fully understand their answers before attempting the practice assessment questions.

Worked example 1

The diagram shows a mass attached by a piece of string to a glider, which is free to glide along an air track.

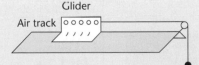

Fig 1.29

A student finds that the glider takes 1.13 s to move a distance of 90 cm starting from rest.

Show that the speed of the glider after 1.13 s is approximately 1.6 m s^{-1}.   **[3]**
Calculate its average acceleration during this time.   **[3]**
How would you test whether or not the acceleration of the glider is constant?   **[3]**

                                 (Total 9 marks)
              (*Edexcel Module Test PH1, January 1997, Q. 5*)

In all answers, each tick indicates the awarding of a single mark – note that no half-marks are ever awarded.

Helpful hint

Always give your answer to a 'show that' question to at least one more significant figure than the value asked for in the question.

Helpful hint

You often find that a 'show that' value is needed later on in the same question. Either use the value provided or your own if you are happy with it.

Answer: First find the average speed of the glider over the distance of 90 cm.

$$\text{average speed} = \text{distance/time} = (0.90 \text{ m})/(1.13 \text{ s}) \quad \checkmark$$

Since the glider starts from rest

$$\text{final speed} = 2 \times \text{average speed} \quad \checkmark$$
$$= 2 \times (0.90 \text{ m})/(1.13 \text{ s}) = 1.59 \text{ m s}^{-1} \quad \checkmark$$

$$\text{average acceleration} = \text{change in velocity/time taken} \quad \checkmark$$
$$= (1.59 \text{ m s}^{-1})/(1.13 \text{ s}) \quad \checkmark$$
$$= 1.41 \text{ m s}^{-2} \quad \checkmark$$

To answer the last part, you can briefly refer to any of the methods described in Experiment 2: measuring acceleration. However your method must compare values of acceleration as the glider moves along the air track. So, if using one of the light gate timing methods, it is important to say that the experiment has to be repeated with the light gate(s) placed at different positions along the air track.

For all methods the mark scheme will be similar:
     selection of correct apparatus
     collection and processing of correct results
     how results used to show acceleration is constant.

For example: if you describe a tickertape method:
     glider pulls a length of tickertape through a ticker-timer $\checkmark$
     cut tape up into equal time lengths and place side by side $\checkmark$
     if constant acceleration, chart will show a constant rate of rise. $\checkmark$

Worked example 2

State the difference between scalar and vector quantities.   **[2]**
A lamp is suspended from two wires as shown in the diagram. The tension in each wire is 4.5 N.

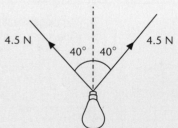

Fig 1.30

Helpful hint

This question starts with a couple of easy marks provided you've remembered to include the 'General Requirements' in your revision!

Calculate the magnitude of the resultant force exerted on the lamp by the wires.   **[3]**
What is the weight of the lamp? Explain your answer.   **[2]**
         (Total 7 marks)
(Edexcel Module Test PH1, January 1999, Q. 3)

Answer:
     scalar quantities do not include a direction $\checkmark$
     vector quantities do include a direction $\checkmark$

Split each tension into a horizontal and a vertical component. The two horizontal components act in opposite directions and cancel. So the resultant force on the lamp is the sum of the two upward vertical components.

vertical component of one tension = 4.5 N × cos 40° ✓

resultant force = 2 × 4.5 N × cos 40° ✓

= 6.9 N (vertically upwards). ✓

The two wires support the lamp and so no resultant force acts on it.

weight of lamp = 6.9 N ✓

as it is in equilibrium (vertical force down = vertical force up). ✓

Worked example 3

The diagram shows part of a roller coaster ride. In practice, friction and air resistance will have a significant effect on the motion of the vehicle, but you should ignore them throughout this question.

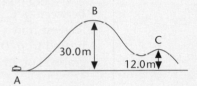

Fig 1.31

The vehicle starts from rest at A and is hauled up to B by a motor. It takes 15.0 s to reach B, at which point its speed is negligible. Complete the energy conversion shown below for the journey from A to B.

Useful work done by motor → .... **[1]**

The mass of the vehicle and the passengers is 3400 kg. Calculate

(i) the useful work done by the motor
(ii) the power output of the motor. **[4]**

At point B the motor is switched off and the vehicle moves under gravity for the rest of the ride. Describe the overall energy conversion which occurs as it travels from B to C. **[1]**

Calculate the speed of the vehicle at point C. **[3]**

On another occasion there are fewer passengers in the vehicle; hence its total mass is less than before. Its speed is again negligible at B. State with a reason how, if at all, you would expect the speed at C to differ from your previous answer. **[2]**

(Total 11 marks)
(Edexcel Unit Test PHY1, June 2001, Q. 5)

Answer: Useful work done by motor → gravitational potential energy. ✓

(i) useful work done by motor = $mg\Delta h$ = 3400 kg × 9.81 N kg^{-1} × 30 m ✓
= 1.00 MJ ✓

(ii) power output = work done/time = 1.00 MJ/(15.0 s) ✓
= 67 kW ✓

Overall energy conversion from B to C:
gravitational potential energy → kinetic energy. ✓

Unit 1

$\frac{1}{2} mv^2$ at C = $mg\Delta h$ from B to C where $\Delta h = 30.0$ m $- 12.0$ m $= 18.0$ m ✓

$\frac{1}{2} v^2 = 9.81$ m s^{-2} × 18.0 m ✓

$v = \sqrt{(353 \text{ m}^2 \text{ s}^{-2})} = 18.8$ m s^{-1} ✓

speed will be the same ✓
since both gravitational potential energy and kinetic energy are proportional to m so it cancels out. ✓

Helpful hint

Answers to these questions, together with explanations, are in the Answers section which follows Chapter 6.

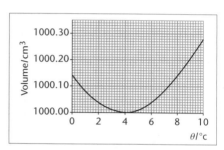

Fig 1.32

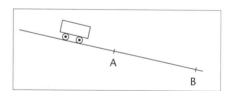

Fig 1.33

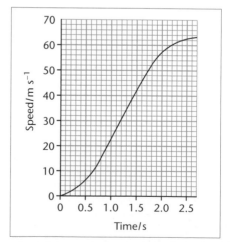

Fig 1.34

Practice questions

The following are typical assessment questions on mechanics. Attempt these questions under similar conditions to those in which you will sit your actual test.

1   The graph in Figure 1.32 shows how the volume of 1.000 kg of water varies with temperature.
State the temperature at which the density of water is a maximum.
[1]

Sketch a graph of how the density of water varies with temperature between 0 °C and 10 °C.
[2]
Suggest how you could demonstrate that the volume of water when heated from 0 °C to 10 °C behaves in the manner indicated by the graph. You may be awarded a mark for the clarity of your answer.
[4]
(Total 7 marks)
(Edexcel Unit Test PHY1, June 2001, Q. 6)

2   Figure 1.33 shows a toy truck, about 30 cm long, accelerating freely down a gentle incline.
Explain carefully how you would measure the average speed with which the truck passes the point A.
[4]
You find that the measured average speed of the truck is 1.52 m s^{-1} when it passes the point A and 1.64 m s^{-1} when it passes the point B. The distance from A to B is 1.20 m. Calculate the acceleration of the truck.
[2]
(Total 6 marks)
(Edexcel Module Test PH1, June 1999, Q. 3)

3   The graph in Figure 1.34 shows the speed of a racing car during the first 2.6 s of a race as it accelerates from rest along a straight line.

Use the graph to estimate

(i)   the displacement 1.5 s after the start   **[2]**
(ii) the acceleration at 2.0 s   **[2]**
(iii) the kinetic energy after 2.5 s given that the mass of the racing car is 420 kg.
[2]
(Total 6 marks)
(Edexcel Module Test PH1, June 1999, Q. 2)

Fig 1.35 A

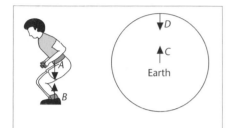

Fig 1.35 B

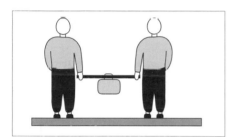

Fig 1.36 A

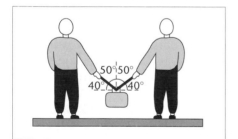

Fig 1.36 B

4 Figure 1.35A shows a child crouching at rest on the ground.

Free-body force diagrams for the child and the Earth are shown in Figure 1.35B.

Complete the following table describing the forces *A*, *B* and *C*.　**[4]**

	Description of force	Body which exerts force	Body the force acts on
Force *A*	Gravitational	Earth	Child
Force *B*			
Force *C*			

All the forces *A*, *B*, *C* and *D* are of equal magnitude.

Why are forces *A* and *B* equal in magnitude?

Why must forces *B* and *D* be equal in magnitude?　**[2]**

The child now jumps vertically upwards. With reference to the forces shown, explain what he must do to jump, and why he then moves upwards.　**[3]**

(Total 9 marks)

(Edexcel Unit Test PHY1, June 2001, Q. 3)

5 Two campers have to carry a heavy container of water between them. One way to make this easier is to pass a pole through the handle as shown in Figure 1.36A.

The container weighs 400 N and the weight of the pole may be neglected. What force must each person apply?　**[1]**

An alternative method is for each person to hold a rope tied to the handle as shown in Figure 1.36B.

Draw a free-body force diagram for the container when held by the ropes.　**[2]**

The weight of the container is 400 N and the two ropes are at 40° to the horizontal. Show that the force each rope applies to the container is about 300 N.　**[3]**

Suggest **two** reasons why the first method of carrying the container is easier.　**[2]**

Two campers using the rope method find that the container keeps bumping on the ground. A bystander suggests that they move further apart so that the ropes are more nearly horizontal. Explain why this would not be a sensible solution to the problem.　**[1]**

(Total 9 marks)

(Edexcel Unit Test PHY1, June 2001, Q. 4)

6 Define linear momentum.　**[1]**

The principle of conservation of linear momentum is a consequence of Newton's laws of motion. An examination candidate is asked to explain this, using a collision between two trolleys as an example. He gives the following answer, which is correct but incomplete. The lines of his answer are numbered on the left for reference.

1　During the collision the trolleys push each other.

2　These forces are of the same size but in opposite directions.

3　As a result, the momentum of one trolley must increase at the same rate as the momentum of the other decreases.

4　Therefore the total momentum of the two trolleys must remain constant.

In which line of his argument is the candidate using Newton's second law?　**[1]**

In which line is he using Newton's third law? **[1]**

The student is making one important assumption which he has not stated. State this assumption. Explain at what point it comes into the argument. **[2]**

Describe how you would check experimentally that momentum is conserved in a collision between two trolleys. **[4]**

(Total 9 marks)

(Edexcel Unit Test PHY1, January 2001, Q. 3)

7 A car travelling at 30 m s⁻¹ collides with a wall. The driver, wearing a seatbelt, is brought to rest in 0.070 s.

The driver has a mass of 50 kg. Calculate the momentum of the driver before the crash. **[2]**

Calculate the average resultant force on the driver during impact. **[3]**

Explain why the resultant force is not the same as the force exerted on the driver by the seatbelt. **[1]**

(Total 6 marks)

(Edexcel Unit Test PHY1, June 2001, Q. 1)

8 The 'London Eye' is a large wheel which rotates at a slow steady speed in a vertical plane about a fixed horizontal axis. A total of 800 passengers can ride in 32 capsules equally spaced around the rim. A simplified diagram is shown in Figure 1.37.

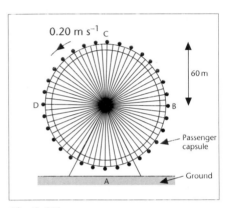

Fig 1.37

On the wheel, the passengers travel at a speed of about 0.20 m s⁻¹ round a circle of radius 60 m. Calculate how long the wheel takes to make one complete revolution. **[2]**

What is the change in the velocity of a passenger when he travels from point B to point D? **[2]**

When one particular passenger ascends from point A to point C, his gravitational potential energy increases by 80 kJ. Calculate his mass. **[3]**

Sketch a graph showing how the gravitational potential energy of this passenger would vary as he ascended from A to C. Add a scale to each axis. **[3]**

Discuss whether it is necessary for the motor driving the wheel to supply this gravitational potential energy. **[2]**

(Total 12 marks)

(Edexcel Unit Test PHY1, January 2001, Q. 2)

Part ② Radioactivity

 Introduction

Radioactivity is the spontaneous emission of particles and electromagnetic waves from the atomic nucleus of certain elements. It was first discovered in 1896 and soon used in the scattering experiment that led to the development of the nuclear model of an atom, in a similar way to that in which electrons are used today to reveal the quark structure of protons and neutrons. As you study radioactivity, you learn about alpha, beta and gamma emissions, their properties and the way in which they each alter their parent nucleus. You find that despite radioactivity being a random process, the rate of decay of a given radioactive material follows a predictable pattern.

 Things to understand

The nuclear atom

- the structure of an atom was discovered by scattering alpha particles from gold foil
- an atom consists of a very small, central nucleus, containing almost all the atom's mass, around which electrons orbit
- an atom is neutral: the nucleus is positive, electrons are negative
- a nucleus consists of a mixture of particles known as nucleons, where a nucleon is either a proton (positive) or a neutron (neutral)
- a nuclear atom is often represented by its nuclear symbol (Figure 1.38) from which the numbers of protons, neutrons and orbiting electrons can be determined
- atoms with the same number of protons in their nuclei can have different numbers of neutrons and so form different isotopes of the same element
- both protons and neutrons are now known to have their own substructure of particles known as quarks
- the quark structure of a nucleus can be revealed by scattering experiments using high energy electrons

$$^{79}_{35}\text{Br}$$

Fig 1.38 *An atom of bromine-79 has a nucleus containing 35 protons and 44 neutrons (a total of 79 nucleons) surrounded by 35 orbiting electrons*

Radiations released during radioactive decay

- all emissions from radioactive decay come from the nucleus
- alpha (positive), beta (usually negative but can be positive) and gamma (no charge) radiations are emitted by a variety of nuclei
- alpha radiation produces a lot of ionisations as the alpha particles push their way through a material – consequently alpha radiation soon runs out of energy and has a very short range

$$^{208}_{84}\text{Po} \longrightarrow {}^{204}_{82}\text{Pb} + {}^{4}_{2}\alpha$$

Fig 1.39 *Polonium-208 decays by alpha emission into lead-204*

(a)

$$^{1}_{0}\text{n} \longrightarrow {}^{1}_{1}\text{p} + {}^{0}_{-1}\text{e} \rangle \beta^{-}$$

(b)

$$^{66}_{29}\text{Cu} \longrightarrow {}^{66}_{30}\text{Zn} + {}^{0}_{-1}\beta$$

Fig 1.40 *(a) A neutron in the nucleus decays into a proton and an electron; the electron is then emitted as a negative beta particle (b) Copper-66 decays by beta-minus emission into zinc-66*

$$^{23}_{12}\text{Mg} \longrightarrow {}^{23}_{11}\text{Na} + {}^{0}_{1}\beta$$

Fig 1.41 *Magnesium-23 decays by beta-plus emission into sodium-23*

- alpha particles are helium nuclei and alpha decay removes two protons and two neutrons from the parent nucleus (Figure 1.39)
- beta radiation produces fewer ionisations and so its particles can travel further than alpha particles before running out of energy
- a beta-minus particle is an electron produced when a neutron in a nucleus splits up into a proton and an ejected 'beta' electron (Figure 1.40)
- a beta-plus particle is a positron (a positive electron) produced when a proton in a nucleus changes into a neutron and an ejected 'beta' positron (Figure 1.41)
- gamma radiation produces very few ionisations along its path and so has a very large range
- gamma radiation is an electromagnetic wave that takes away any surplus energy that a nucleus may have been left with after it has emitted either alpha or beta particles

Radioactive decay rates

- all radioactive decay is random; the time at which a particular nucleus will decay is unpredictable
- the activity of a source depends on the total number of nuclei present at that time
- the activity of a source decreases with time as the decays taking place reduce the number of nuclei left to decay
- an activity–time graph produces an exponential decay curve
- the average time taken for the activity to drop to half its value (the half-life) is the same throughout a given decay but varies from source to source
- all activity measurements should be adjusted to remove the background activity produced by naturally occurring radio-isotopes and cosmic rays

 Things to learn

You should learn the following for your Unit PHY1 Test. Remember that it may also test your understanding of the 'general requirements' (see Appendix 1).

Equations

All radioactivity equations are provided but you do need to learn the nuclear symbols for the following particles to complete radioactive decay equations:

- ❑ alpha particle $\qquad$ $^{4}_{2}\alpha$ or $^{4}_{2}\text{He}$
- ❑ beta-minus particle $\qquad$ $^{0}_{-1}\beta$ or $^{0}_{-1}\text{e}^{-}$
- ❑ beta-plus particle $\qquad$ $^{0}_{1}\beta$ or $^{0}_{1}\text{e}^{+}$
- ❑ gamma radiation $\qquad$ $^{0}_{0}\gamma$ or γ
- ❑ neutron $\qquad$ $^{1}_{0}\text{n}$
- ❑ proton $\qquad$ $^{1}_{1}\text{p}$

General definitions

- ❑ nucleus: very small, positive centre of an atom in which nearly all the atom's mass is concentrated

- nucleons: protons and neutrons – the basic particles from which the nucleus of an atom is constructed
- quarks: the basic particles from which protons, neutrons and many other sub-atomic particles are constructed
- isotopes: atoms that have the same number of protons but a different number of neutrons in their nuclei
- background radiation: random emissions from naturally occurring radio-isotopes that must be taken into account whenever performing radioactivity experiments
- activity: the number of nuclei of a source that decay in one second
- becquerel (Bq): a unit of activity; a count rate of one disintegration per second
- decay constant: the proportion of the nuclei present that decay in one second
- half-life: the average time taken for half the nuclei of that radioactive element to decay *or* the average time for the activity to fall to 50% of its original value

Word equation definitions

Use the following word equation when asked to define:

- decay constant = activity/number of nuclei present

Experiments

Although you will not have performed all these experiments yourself, you may still be asked to describe them!

1. Alpha particle scattering experiment

Alpha particles are fired at thin gold foil (Figure 1.42).
The coated screen flashes when an alpha particle hits it.
Most of the alpha particles pass almost straight through the foil.
Some alpha particles deflect through small angles.
A very small minority of alpha particles (about 1 in 8000) deflects through more than 90°.
Conclusions:
1 an atom has a very tiny charged centre (the nucleus), containing most of the atom's mass
2 the nuclei have comparatively large distances between them.

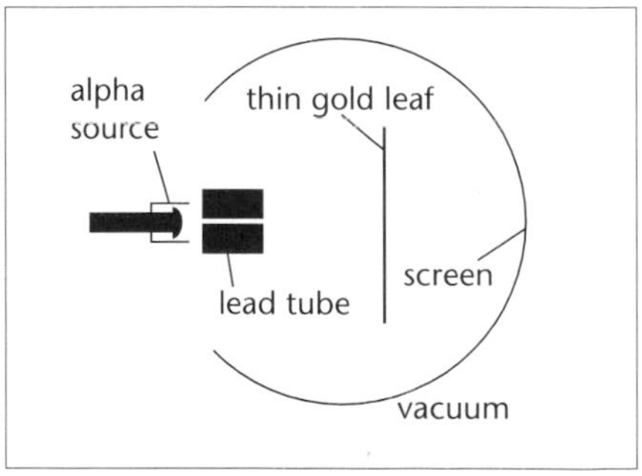

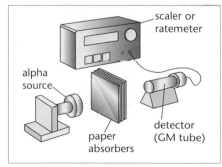

Fig 1.42 *The alpha particle scattering experiment*

2. Penetration of the radiations emitted by radioactive sources

Use a Geiger–Muller (GM) tube with a thin window so that alpha particles can pass into it and so be detected.
Record a number of count rates with no source present and obtain an average background count.
Keep each source a fixed distance from the GM tube (e.g. 1 cm for the alpha source, 3 cm for the beta source and 6 cm for the gamma source).
Measure the corrected count rate for the alpha source for different thicknesses of paper between it and the GM tube (Figure 1.43).
Repeat for the beta source using thin pieces of aluminium as the absorber.
Repeat for the gamma source using different thicknesses of lead absorbers.

Fig 1.43 *Thin paper stops alpha particles*

Results:
1 alpha particles are stopped by thin paper
2 beta particles can penetrate up to several millimetres of aluminium
3 gamma radiation can still be detected after passing through several centimetres of lead.

❏ 3. Measuring the half-life of protactinium-234

Record a number of count rates with no source present and obtain an average background count.

Shake the 'protactinium generator' to transfer the protactinium compound from the lower water-based layer to the upper organic layer.

When the layers re-establish, place the GM tube alongside the top layer (Figure 1.44).

Record the count rate at intervals of 10 s for 5 min.

Plot a graph of corrected count rate against time (Figure 1.45).

From the graph determine how long it takes for the count rate at any given time to halve its value.

Protactinium-234 is used as it has a half-life of about 1 min and therefore its activity falls significantly during a 5-min period.

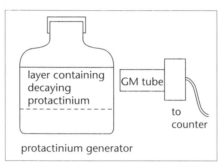

Fig 1.44 *The GM tube monitors the decay of protactinium in the top layer*

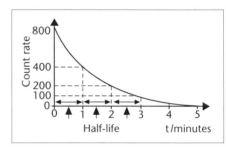

Fig 1.45 *The half-life of this isotope is about 1 min*

 Checklist

Before attempting the following questions on radioactivity, check that you:

❏ have learnt a description of the alpha particle scattering experiment and know how its results led to the nuclear model of an atom

❏ appreciate that the diameter of an atom is about 10^{-10} m and that of a nucleus is about 10^{-15} m

❏ know the structure of a nucleus and how to find the number of protons and neutrons it contains from its nuclear symbol

❏ understand the term 'isotope'

❏ can compare the similarities and differences between the alpha particle scattering experiment and deep inelastic scattering of electrons

❏ appreciate that both protons and neutrons have a sub-structure consisting of three quarks

❏ know the nature of the radiations emitted by a radioactive source

❏ have learnt the nuclear symbols for an alpha particle, both types of beta particle, gamma radiation, a neutron and a proton

❏ can complete and balance nuclear equations

❏ know how to distinguish experimentally between alpha, beta and gamma radiations with reference to their ranges in air and their penetrations through different absorbers

- appreciate the link between a radiation's ionising ability and its penetrating power *or* range
- know some sources of background radiation
- appreciate that the random and unpredictable nature of an individual decay still leads to an overall predictable decay pattern for the source as a whole
- know the meanings and units of 'activity', 'decay constant' and 'half-life'
- can use the equations that relate these quantities
- have learnt a description of an experiment to measure the half-life of a radioisotope with a half-life of about a minute
- are familiar with the 'general requirements' (see Appendix 1) and how they apply to the topic of radioactivity

 Testing your knowledge and understanding

Quick test

Answers to these questions, together with explanations, are in the Answers section which follows Chapter 6.

Select the correct answer to each of the following questions from the four answers supplied. In each case only one of the four answers is correct. Allow about 40 min for the 20 questions.

1 Experiments in which beams of alpha particles were fired at thin metal foils provided evidence for the existence of

    A quarks   **B** isotopes   **C** positrons   **D** nuclear atoms

2 Three of the following statements about a neutral atom of an element are correct. Which is the incorrect statement?

    A The number of electrons in the atom equals the number of protons in the nucleus
    B The proton number is the same for all atoms of all isotopes of the same element
    C The proton number is the nearest integer to the mass number
    D Neutron number = nucleon number – proton number

3 The nuclear symbol for an isotope of bismuth is $^{209}_{83}$Bi. An atom of this isotope consists of

    A 83 protons, 43 neutrons, 83 electrons
    B 83 protons, 126 neutrons, 83 electrons
    C 83 protons, 209 neutrons, 209 electrons
    D 126 protons, 83 neutrons, 126 electrons

4 $^{4}_{2}$He represents the helium nucleus. Three of the following statements about the helium nucleus are correct. Which is the incorrect statement?

    A It contains two neutrons
    B It contains two protons
    C It has to gain four orbital electrons to become a helium atom
    D It is called an alpha particle when emitted by a radioactive source

5 A proton consists of

    A a mixture of nucleons
    B a neutron and an electron
    C a mixture of quarks
    D a nucleus and an electron

6 The radiation from a radioactive source is found to pass through a sheet of paper without any reduction in its intensity but is completely absorbed by a piece of aluminium 1 cm thick. The radiation consists of

    A alpha radiation only

    B beta radiation only

    C a mixture of alpha and beta radiations

    D a mixture of alpha, beta and gamma radiations

7 When a Geiger counter is brought near to a closed lead box containing a radioactive source, the measured count rate rises from 0.1 to 0.9 Bq. The increase in the count rate is due to

    A background radiation + beta and gamma radiation from the source

    B background radiation + gamma radiation from the source

    C beta and gamma radiation from the source

    D gamma radiation from the source

8 An isotope of radon, $^{220}_{86}\text{Rn}$, decays by emitting alpha radiation. Which of the following represents the product of this alpha decay?

    A $^{224}_{88}\text{Ra}$ **B** $^{222}_{90}\text{Th}$ **C** $^{216}_{84}\text{Po}$ **D** $^{218}_{82}\text{Pb}$

9 When an unstable radioactive isotope X emits beta-minus radiation, a stable isotope Y is formed. The decay can be represented by the nuclear equation $X \rightarrow Y + \beta^-$. Which of the following statements is correct?

    A X has a smaller proton number than Y

    B X and Y have the same proton number

    C X and Y are isotopes of the same element

    D X and Y have different nucleon numbers

10 After the nucleus of a cobalt atom, $^{60}_{27}\text{Co}$, has emitted gamma radiation the number of protons in the resulting nucleus is

    A 25   **B** 27   **C** 33   **D** 60

11 A radioactive nuclide of gold, $^{197}_{79}\text{Au}$, decays to form a nuclide of platinum, $^{197}_{78}\text{Pt}$. The gold decayed by emitting

    A an alpha particle

    B a beta-minus particle

    C a beta-plus particle

    D gamma radiation

12 The boron isotope $^{10}_{5}\text{B}$ reacts with another particle to produce the lithium isotope $^{7}_{3}\text{Li}$ and an alpha particle. The other particle is

    A an electron   **B** a neutron   **C** a positron   **D** a proton

13 A radioactive isotope of element P has a proton number Z and a nucleon number A. It emits an alpha particle to become element Q which then emits a negative beta particle to become element R. What is the proton number and the nucleon number of element R?

    A $Z - 1$ and $A - 2$

    B $Z - 4$ and $A - 2$

    C $Z - 1$ and $A - 4$

    D $Z - 4$ and $A - 4$

14 Which of the following radioactive disintegrations will result in the formation of a different isotope of the parent substance?

    A Gamma radiation

    B An alpha particle + a beta particle

    C An alpha particle + two beta particles

    D Two alpha particles + a beta particle

15 A sample of the radioactive gas, radon, is placed in a container where it causes some of the air to ionise. The amount of ionisation is found to decrease with time. The decrease is most likely due to

  A a decrease in the number of radon gas atoms present
  B a decrease in the number of air atoms present
  C mixing of the radon gas with the air
  D a chemical reaction between the radon gas and the air

16 A radioactive element emits alpha particles. It has a half-life of 10 days. What will be the mass of this element remaining after 40 days in a sample initially containing 48 g?

  A 3.0 g   **B** 6.0 g   **C** 12 g   **D** 48 g

17 The radioactive element $^{226}_{88}$Ra has a half-life of about 1600 years. An old sample of $^{226}_{88}$Ra is observed at a certain time to be emitting radiation at approximately 100 counts per minute. At what approximate rate, in counts per minute, would the sample have been emitting radiation 16 years previously?

  A 100   **B** 160   **C** 1600   **D** 10 000

18 Radon has a half-life of 50 s. Three of the following statements about a sample of 8 g of radon are correct. Which is the incorrect statement?

  A The count rate will have fallen to about $\frac{1}{4}$ of its initial value after 100 s
  B There will be about 1 g of undecayed radon left after 150 s
  C The count rate will have fallen to about 1/16 of its initial value after 200 s
  D There will be about $\frac{1}{4}$ g of undecayed radon left after 200 s

19 In order to trace the line of a water-pipe buried 0.4 m below the surface of a field, an engineer wishes to add a radioactive isotope to the water. Which isotope should he choose?

  A A beta emitter with a half-life of a few hours
  B A beta emitter with a half-life of several years
  C A gamma emitter with a half-life of a few hours
  D A gamma emitter with a half-life of a several years

20 The decay rate for a sample of radon gas at a particular instant is 7560 Bq. The decay constant for radon is 0.0126 s^{-1} and its mass number is 220. The number of radon nuclei present in the sample at this particular instant is

  A 95   **B** 17 460   **C** 600 000   **D** 1 660 000

Worked examples

Study the following worked examples on radioactivity carefully. Make sure you fully understand their answers before attempting the practice assessment questions.

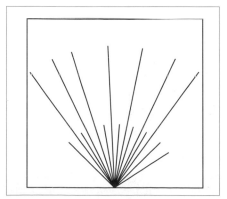

Fig 1.46

Worked example 1

Figure 1.46 shows a diagram of the tracks produced by alpha particles emitted by a radioactive source.

Suggest properties of the alpha particles that can be deduced from this diagram?

[4]

Helpful hint

Link your knowledge of alpha particles with details in the photograph. As with most 'suggest' questions, there are a number of ways of achieving the maximum 4 marks.

Helpful hint

You cannot use an absorber to stop the beta particles as this will also stop all the alpha particles. You have to stop the particles that you are trying to count!

A different source emits both alpha and beta particles. How would you use a Geiger counter to determine the approximate count rate due to the alpha radiation only? **[3]**

(Total 7 marks)

(*Edexcel Module Test PH2, June 1999, Q. 2*)

Answer:

(dense tracks)	alpha particles are strongly ionising/are charged ✓ and, consequently, will have a short range ✓
(straight tracks)	alpha particles travel in straight lines/have large mass/momentum ✓
(same length tracks)	alpha particles are emitted with the same energy/speed *or* have the same range ✓
(two track lengths)	source emits particles with two distinct energies. ✓ **Max 4**

For example, find count rate using a thin window GM tube placed close to the source ✓
(this will be the combined count rate for the alpha and beta particles)
find count rate with thick paper between GM tube and source ✓
(this will be the count rate for the beta particles only)
alpha count rate = difference in these two readings. ✓

Worked example 2

Tritium ^3_1T decays by beta-minus emission. ^3_1T has a half-life of 12 years.
Of which element is ^3_1T an isotope? **[1]**
Complete the following nuclear equation for this decay.
$^3_1T \rightarrow X +$ **[2]**
Define the term *half-life*. **[2]**

(Total 5 marks)

(*Edexcel Module Test PH2, January 1998, Q. 1*)

Answer:
Proton number of tritium = 1
so tritium is an isotope of hydrogen. ✓

Beta-minus particle = $^{\ 0}_{-1}\beta$
so $^3_1T \rightarrow {}^3_2X + {}^{\ 0}_{-1}\beta$ ✓✓

Half-life: the average time taken ✓
for half the nuclei of that radioactive element to decay. ✓

Worked example 3

It is thought that an extremely short-lived radioactive isotope $^{269}_{110}X$, which decays by alpha emission, has a half-life of 200 µs. After a series of decays the element $^A_{104}Y$ is formed from the original isotope. There are no beta decays. Deduce the value of A. **[3]**
Show that the decay constant of $^{269}_{110}X$ is approximately 3500 s^{-1}. **[2]**
The number of nuclei N of $^{269}_{110}X$ in a sample of mass 0.54 µg is 1.2×10^{15}.
Determine the activity of 0.54 µg of $^{269}_{110}X$. **[2]**
Why is this value for the activity only approximate? **[1]**

(Total 8 marks)

(*Edexcel Module Test PH2, January 1999, Q. 2*)

Answer: Each alpha decay decreases the nucleon number by 4 and the proton number by 2 ✓

proton number has decreased by 110 − 104 = 6 requiring 3 alpha decays ✓
nucleon number will decrease by 3 × 4 = 12
so A = 269 − 12 = 257 ✓

decay constant $\lambda = 0.69/t_{\frac{1}{2}}$ (no credit as equation provided in exam)
$$= 0.69/(200 \times 10^{-6} \text{ s}) \checkmark$$
$$= 3450 \text{ s}^{-1} \qquad\qquad\qquad\qquad \checkmark$$
$$\approx 3500 \text{ s}^{-1} \text{ as required.}$$

Activity $= \lambda N$ (no credit as equation provided in exam)
$$= 3450 \text{ s}^{-1} \times 1.2 \times 10^{15} \checkmark$$
$$= 4.1 \times 10^{18} \text{ Bq} \qquad (\text{or s}^{-1}) \checkmark$$

Either
since radioactivity is a random process, activity fluctuates from one moment to the next making it impossible to calculate the exact activity at a given instant (*or* idea of half-life being an average value) ✓
or
a half-life of 200 μs is so short that it is very difficult to measure and would itself only be an approximate value ✓

Helpful hint

Make sure you are familiar with the layout and contents of the formulae sheet. Don't waste precious exam time hunting for an equation that might not even be there!

Helpful hint

Answers to these questions, together with explanations, are in the Answers section which follows Chapter 6.

Practice questions

The following are typical assessment questions on radioactivity. Attempt these questions under similar conditions to those in which you will sit your actual test.

1 In 1909 Geiger and Marsden carried out an important experiment to investigate alpha particle scattering. Alpha particles were directed towards a thin gold sheet and detectors were used to observe the distribution of scattered alpha particles. State what was observed in this experiment. **[3]**
Explain why these observations led to the conclusion that an atom was composed mainly of space, with a very small, relatively massive, charged nucleus. **[3]**
State an approximate value for the diameter of (i) a gold atom and (ii) a gold nucleus. **[2]**
(Total 8 marks)
(*Edexcel Module Test PH2, June 1996, Q. 1*)

2 Complete the following table which compares alpha particle scattering and deep inelastic scattering experiments. **[2]**

	Alpha particle scattering	**Deep inelastic scattering**
Incident particles	Alpha particles	
Target		Nucleons

Write a short paragraph describing the conclusion from each experiment. **[2, 2]**
(Total 6 marks)
(*Edexcel Unit Test PHY1, January 2001, Q. 7*)

3 A student has a sample of a radioactive element which is thought to be a pure beta emitter. The student has **only** the following apparatus available:

- a thin window Geiger–Muller (GM) tube connected to a counter
- a piece of aluminium 3 mm thick
- a clock
- a half-metre rule.

How would the student determine the background radiation level in the laboratory? **[2]**

The student investigates how the count rate varies with distance from the source to the GM tube and also the effect of inserting the aluminium absorber. From these experiments explain how the student could confirm that the sample was a pure beta emitter. You may be awarded a mark for the clarity of your answer. **[5]**

(Total 7 marks)

(Edexcel Unit Test PHY1, January 2001, Q. 6)

4 Protactinium, Pa, decays to uranium $^{234}_{92}U$ by emitting a beta-minus particle. The uranium produced is itself radioactive and decays by alpha emission to thorium, Th.

Mark and label the position of $^{234}_{92}U$ on the grid in Figure 1.47. **[1]**

Draw arrows on the grid showing both the beta-minus and the alpha decays referred to above. Label your arrows α and β. **[3]**

(Total 4 marks)

(Edexcel Unit Test PHY1, June 2001, Q. 8)

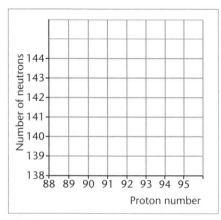

Fig 1.47

5 A student carries out an experiment to determine the half-life of a radioactive isotope M. The student subtracts the mean background count from the readings and plots the smooth curve shown in Figure 1.48.

From this graph the student concludes that the isotope M is not pure, but contains a small proportion of another isotope C with a relatively long half-life. State a feature of the graph that supports this conclusion. **[1]**

Estimate the activity of isotope C. **[1]**

Determine the half-life of isotope M, showing clearly how you obtained your answer. **[3]**

Isotope M decays by beta-minus emission. Write down a nuclear equation showing how the beta-minus particles are produced within the nucleus. **[1]**

Describe briefly how the student could determine the nature of the radiation emitted by isotope C. **[3]**

(Total 9 marks)

(Edexcel Module Test PH2, June 1998, Q. 3)

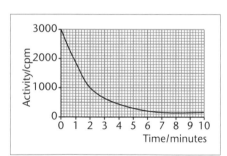

Fig 1.48

6 State the number of protons and the number of neutrons in $^{14}_{6}C$. **[2]**

The mass of one nucleus of $^{14}_{6}C = 2.34 \times 10^{-26}$ kg.

The nucleus of carbon-14 has a radius of 2.70×10^{-15} m.

Show that the volume of a carbon-14 nucleus is about 8×10^{-44} m³. **[2]**

Determine the density of this nucleus. **[2]**

How does your value compare with the densities of everyday materials? **[1]**

Carbon-14 is a radioisotope with a half-life of 5700 years. What is meant by the term half-life? **[2]**

Calculate the decay constant of carbon-14 in s^{-1}. **[2]**

(Total 11 marks)

(Edexcel Unit Test PHY1, January 2001, Q. 4)

7 A student measured the background radiation in a laboratory at 4.0 Bq. State **two** sources of background radiation. **[2]**

Sodium-22 decays by beta-plus radiation to neon. Complete the following nuclear equation for this decay ensuring each symbol has the appropriate nucleon and proton numbers.

$$^{22}_{11}\text{Na} \rightarrow \text{Ne} +$$ **[2]**

Write down another possible isotope of sodium. **[1]**

Sodium-22 has a half-life of 2.6 years. Determine the decay constant of sodium-22 in s^{-1}. **[2]**

A sample of common salt (sodium chloride) is contaminated with sodium-22. The activity of a spoonful is found to be 2.5 Bq. How many nuclei of sodium-22 does the spoonful contain? **[2]**

Explain whether your answer suggests that the salt is **heavily** contaminated. **[1]**

(Total 10 marks)

(Edexcel Unit Test PHY1, June 2001, Q. 7)

2 Electricity and thermal physics

Part 1 Electricity

✳ Introduction

Electricity is that 'invisible' method of working that has transformed the way we live. Modern civilisation relies heavily on electricity. Electricity drives modern life like few other discoveries of the nineteenth century. Among other things, it lights your house, heats the water in your kettle, turns the motor in your washing machine and powers your television and video recorder. Electricity is convenient and easy to control. As you study electricity, you learn how the very slow movement of a large number of electrons within an electric circuit results in a current. You find how the size of a current depends on the potential difference driving the charges and on the resistance to their movement. This allows you to calculate the current in and the power dissipated by different parts of a circuit, even when the power supply has its own internal resistance. You discover how the resistance of a wire depends on its length and cross-sectional area, how the resistance of a light-dependent resistor (LDR) depends on its illumination and how the resistance of a thermistor depends on its temperature.

✳ Things to understand

Charge and current

- charge Q is either positive or negative and is measured in coulombs (C)
- an electric current I consists of a flow of charged particles: these might be positive (a beam of alpha particles), negative (electrons moving through a wire) or a mixture of both positive and negative (ions moving through an electrolyte solution)
- the direction of a current is always that in which a positive charge would move irrespective of the sign of the actual charge carriers (Figure 2.1)
- current is the rate of flow of charge
- current is measured in coulombs per second (C s^{-1}), a unit commonly known as the ampere (A)
- the current flowing at a point in a circuit is measured by inserting an ammeter into the circuit at that point (Figure 2.2)

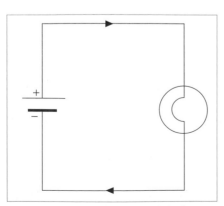

Fig 2.1 *The current direction is from the positive to the negative terminal*

2.1 ELECTRICITY

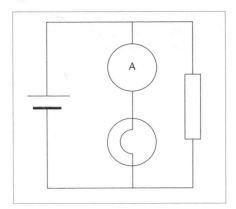

Fig 2.2 *This ammeter is measuring the current in the lamp*

- an ammeter should have as low a resistance as possible
- charge is always conserved
- current is the same at all points around a series circuit
- the total current entering a junction equals the total current leaving it

Drift velocity

- the electrical behaviour of metals, semiconductors and insulators is related to the number of charged particles that are free to move within them
- all electrons in a perfect insulator are fixed to their atoms so that none are free to move
- most electrons in a metal are fixed to their atoms but one or two per atom are free to move and carry charge
- the number of atoms and, hence, the number of conduction electrons in a metal wire is extremely large
- the current in a metal wire consists of a very large number of electrons moving at a very slow drift velocity, typically 0.1 mm s^{-1}
- the equation $I = nAqv$ (Table 2.1 defines the symbols used and gives their units)

symbol	quantity	unit
I	current	A
n	charge carrier density (number of charge carriers per unit volume)	m^{-3}
A	cross-sectional area	m^2
q	charge per carrier	C
v	drift velocity	m s^{-1}

Table 2.1 *Symbols used in the equation I = nAqv*

- when a circuit is turned on, all the conduction electrons throughout the circuit start to slowly drift almost instantaneously

Potential difference

- in a circuit, the power supply forces the charge carriers to move around a circuit from a high to a low potential
- components oppose the motion of these charges
- the moving charges do work on the components and transfer energy to them
- components in which there is a current must have a potential difference across them
- the potential difference across a component is the energy transferred by each coulomb of charge passing through it
- potential difference is measured in joules per coulomb (J C^{-1}), a unit commonly known as the volt (V)
- the potential difference across a component is measured by connecting a voltmeter in parallel with that component (Figure 2.3)
- a voltmeter should have as high a resistance as possible

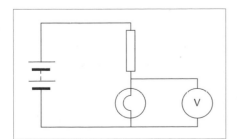

Fig 2.3 *This voltmeter is measuring the potential difference across the lamp*

- components connected in parallel must have the same potential difference across them
- the total potential difference across components connected in series is the sum of the potential differences across each one

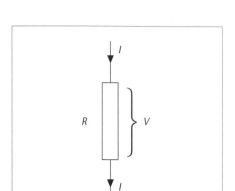

Fig 2.4 $R = V/I$

Resistance

- resistance is a measure of a component's opposition to the flow of charge
- when resistance is high, a large potential difference is needed for a given current
- resistance is the ratio of the potential difference across the component to the current in it (Figure 2.4)
- the resistance of a component indicates the potential difference needed for 1 A of current
- an ohmmeter can be used to measure directly the resistance of a component, although if the component's resistance is small then the resistance of the connecting leads must be taken into account
- connecting components in series increases the total resistance
- the total resistance of components in parallel is always less than any of the individual resistances
- the resistance of a metal wire increases with temperature since the increased lattice vibrations obstruct the movement of the charge carriers and so reduce their drift velocity
- the resistance of a negative temperature coefficient (NTC) thermistor decreases as temperature increases since the increased lattice vibrations release more charge carriers so increasing the charge carrier density
- LDR resistance decreases as illumination increases

Resistivity

- resistivity is a measure of a material's opposition to the flow of charge
- the resistance of a wire depends on its length, cross-sectional area and the resistivity of the material from which it is made
- a long wire has a greater resistance than a short wire of the same material and area
- a thick wire has a smaller resistance than a thin wire of the same material and length

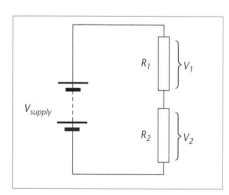

Fig 2.5 $V_1 = V_{supply} \times R_1/(R_1 + R_2)$ and $V_2 = V_{supply} \times R_2/(R_1 + R_2)$ and $V_1/V_2 = R_1/R_2$

Potential divider

- a potential divider consists of a chain of resistors connected in series across a supply voltage
- a potential divider divides the total potential difference across it in the ratio of its resistances so the potential difference across each resistor depends on the values of the resistances used (Figure 2.5)
- a temperature-sensitive potential divider uses a thermistor as one of its series resistors
- a light-sensitive potential divider uses an LDR as one of its series resistors
- a potentiometer is a variable potential divider (Figure 2.6)

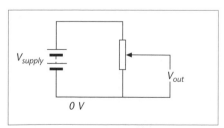

Fig 2.6 Using a potentiometer, V_{out} can be varied continuously from 0 V to V_{supply}

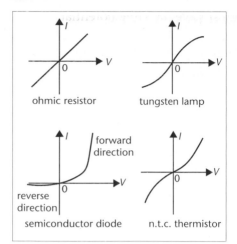

Fig 2.7 *I–V characteristics for different components*

Current–potential difference characteristics

- the way in which the current changes with potential difference varies from component to component (Figure 2.7)
- current is proportional to potential difference for an ohmic component; the graph is a straight line through the origin and the resistance is constant
- the graph for a tungsten filament lamp shows that the resistance of its filament increases as it gets hotter
- the graph for a negative temperature coefficient thermistor shows that its resistance decreases as it gets hotter
- a semiconductor diode only conducts well in the forward direction while very little current flows in the reverse direction

E.m.f. and internal resistance

- a power supply does work on the charge carriers as it forces them to move around a circuit and so gives energy to the circuit
- the e.m.f. of a power supply is the energy given to each coulomb of charge passing through it
- e.m.f. is measured in joules per coulomb ($J C^{-1}$) or volts (V)
- not all the energy given to the charge carriers reaches the external circuit as some of it is transferred within the power supply as the charges do work on the internal resistance
- internal resistance is the effective opposition of the power supply to the flow of current through it
- the potential difference across the internal resistance is often referred to as the 'lost volts'
- the potential difference across the terminals of a power supply only equals its e.m.f. when it is supplying no current
- the terminal potential difference is less than the e.m.f. when there is a current (Figure 2.8); the larger the current, the greater the 'lost volts' and the smaller the terminal potential difference
- a car battery has a very small internal resistance as it has to supply large currents to the starter motor
- an e.h.t. power supply has a very large internal resistance to limit the current it supplies to a safe level

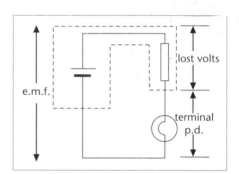

Fig 2.8 *Terminal potential difference = e.m.f. – 'lost volts'*

Circuit calculations

- total circuit resistance must include the internal resistance of the power supply
- current supplied = e.m.f./total circuit resistance
- potential difference across a part of a circuit = current in that part × resistance of that part

 Things to learn

You should learn the following for your Unit PHY2 Test. Remember that it may also test your understanding of the 'general requirements' (see Appendix 1).

Unit 2

Equations that will *not* be given to you in the test

- ❑ charge = current × time
  $$Q = It$$
  Q = charge
  I = current

- ❑ potential difference = energy transferred/charge
  $$V = W/Q$$
  V = potential difference
  W = energy transferred

- ❑ potential difference = current × resistance
  $$V = IR$$
  R = resistance

- ❑ electrical power = potential difference × current
  $$P = VI$$

- ❑ energy transferred electrically = potential difference × current × time
  $$W = VIt$$

- ❑ resistance = resistivity × length/cross-sectional area
  $$R = \rho l/A$$
  ρ = resistivity

Laws

- ❑ the sum of the currents entering a point is equal to the sum of the currents leaving that point (known as Kirchhoff's first law; this law is a consequence of the conservation of charge)

- ❑ around any closed loop, the sum of the e.m.f.s is equal to the sum of the potential differences (known as Kirchhoff's second law; this law is a consequence of the conservation of energy)

- ❑ Ohm's Law: the current in a conductor is directly proportional to the potential difference across it provided the temperature remains constant

General definitions

- ❑ current: the rate of flow of charge

- ❑ series: components connected such that the same current goes through each in turn

- ❑ parallel: components connected across each other such that they each have the same potential difference and the current has a choice of routes

- ❑ drift velocity: the average velocity of the charge carriers through a circuit

- ❑ charge carrier density: the number of charged particles per metre cubed that are free to move and carry current

- ❑ potential difference: a voltage across a component that takes energy away from a charge; the energy transferred by each coulomb of charge passing through the component

- ❑ resistor: a component that opposes the flow of current

- ❑ NTC thermistor: a component whose resistance decreases as the temperature increases

- ❑ LDR: a component whose resistance decreases as the level of illumination increases

- ❑ diode: a component that conducts easily in one direction but not in the other

- ❑ potential divider: a chain of resistors that divides up the voltage from a source in proportion to the resistance values

Unit 2

- [ ] e.m.f. (electromotive force): a voltage that does work on, and gives energy to, a charge; the energy given to each coulomb of charge
- [ ] internal resistance: opposition to the flow of current within the power supply

Word equation definitions

Use the following word equations when asked to define:

- [ ] current = charge flowing/time taken   so that $1\ A = 1\ C\ s^{-1}$
- [ ] potential difference = work done/charge   so that $1\ V = 1\ J\ C^{-1}$

  [alternatively, you could use potential difference = power dissipation/current]

- [ ] resistance = potential difference across/current in
- [ ] resistivity = resistance × cross-sectional area/length

Experiments

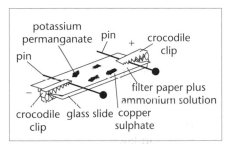

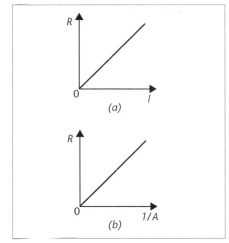

Fig 2.9 *The coloured ions move very slowly*

[ ] 1. Conduction by coloured ions

Wet some filter paper on a glass slide with ammonium solution.
Put a single crystal of copper sulphate and one of potassium permanganate onto the filter paper.
Using pins as electrodes, connect a potential difference across the slide. (Figure 2.9)
The positive blue copper ion moves slowly towards the negative electrode.
The negative purple permanganate ion moves slowly towards the positive electrode.

[ ] 2. The resistance of a wire

Join the terminals of a digital ohmmeter and record the resistance of its connecting leads.
Use the ohmmeter to measure the resistance of different lengths of the same nichrome wire – remember to subtract the resistance of the connecting leads.
Plot a graph of resistance against length (Figure 2.10a).
Use the ohmmeter to measure the resistance of equal lengths of different thicknesses of nichrome wire.
Use a micrometer (having checked for zero error) to measure the diameters.
Plot a graph of resistance against 1/cross-sectional area (Figure 2.10b).
The results show that $R \propto l/A$.
The constant of proportionality is ρ, the resistivity of nichrome, so $R = \rho l/A$.
Use $\rho = RA/l$ to calculate the resistivity of nichrome.

Fig 2.10 *(a) $R \propto l$ (b) $R \propto 1/A$*

[ ] 3. Measuring how current varies with potential difference

Use a range of components: ohmic resistor, lamp, thermistor and diode.
Use a variable power supply to vary the potential difference across the component (Figure 2.11).
Series resistor is needed to prevent damage to diode when in forward direction.
Record a range of corresponding values of current I and potential difference V.
Reverse the component and repeat.
Plot a graph of I against V (Figure 2.7).
Calculate the resistance (V/I) for each pair of values and compare.

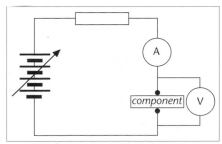

Fig 2.11 *Measure the current for a range of potential differences for each component*

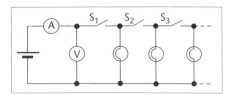

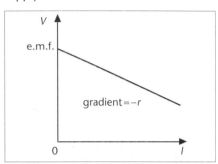

Fig 2.12 The current increases as more lamps are connected across the supply

Fig 2.13 The terminal potential difference decreases as the current increases

❏ 4. Measuring e.m.f. and internal resistance

Set up the circuit shown in Figure 2.12.

Open all switches and record the reading of the digital voltmeter. This is the e.m.f..

Close switch S_1 and record the readings of the ammeter and the voltmeter. Then also close S_2 and record the readings of the ammeter and the voltmeter.

Continue until all switches are closed.

Plot a graph of potential difference against current (Figure 2.13).

Terminal potential difference = e.m.f. – potential difference across internal resistance

$$= \text{e.m.f.} - Ir = -rI + \text{e.m.f.}$$

Comparing this equation with $y = mx + c$
gradient of graph $= -r$
so internal resistance $= -$gradient
intercept $=$ e.m.f.

✳ *Checklist*

Before attempting the following questions on electricity, check that you:

❏ know that the charge on an electron is negative and also very small

❏ appreciate that a very large number of electrons are needed to give one coulomb of charge

❏ realise that any flow of charge (positive, negative or even a mixture of both in opposite directions) is a current

❏ know that the size of a current gives the rate of flow of charge and that $1\,\text{A} = 1\,\text{C s}^{-1}$

❏ understand why an ammeter has a very low resistance and how it is connected to measure a current

❏ know that current is the same at all points in a series circuit and that the sum of the currents entering and leaving a junction is the same

❏ know the meaning and the unit of each symbol in the equation $I = nAqv$

❏ can relate the electrical behaviour of conductors, semiconductors and insulators to the value of 'n'

❏ know that the drift velocity v of an electron in a metal is very slow (typically $0.1\,\text{mm s}^{-1}$)

❏ have learnt a definition of potential difference and know that $1\,\text{V} = 1\,\text{J C}^{-1}$

❏ understand why a voltmeter has a very high resistance and how it is connected to measure potential difference

❏ know that components in parallel have the same potential difference across them

❏ know that the potential differences add up for components in series

❏ can confidently solve problems relating to series and parallel circuits

❏ appreciate the effects that non-ideal ammeters and voltmeters have on a circuit

❏ can calculate, given its resistance, the reading of a non-ideal meter placed anywhere in a circuit

❏ know the effect of resistance on current flow

❑ have learnt the definition of resistance and know its unit

❑ can calculate the effective resistance of series and parallel combinations of resistors

❑ appreciate how an increase in temperature increases a metal's resistance by reducing the charge carriers' drift velocity

❑ appreciate how an increase in temperature reduces a negative temperature coefficient thermistor's resistance by increasing the charge carrier density

❑ know that increased light level decreases the resistance of a light dependent resistor

❑ have measured the resistance of various wires and know how its value depends on the wire's dimensions

❑ have learnt the definition of resistivity and know its unit

❑ can calculate the potential difference across each of the resistors in a potential divider circuit

❑ appreciate that thermistors and light dependent resistors can be used in potential divider circuits to produce temperature and light sensors

❑ know the structure of a modern potentiometer and how it can be used either as a variable resistor to control current or as a potential divider to control voltage

❑ have learnt a description of an experiment to measure the current-potential difference characteristics of a component

❑ can sketch the $I-V$ graphs for:

> an ohmic conductor
> a tungsten (metal) filament lamp
> an NTC thermistor (or carbon filament lamp)
> a semiconductor diode

❑ have learnt a statement of Ohm's law

❑ have learnt the definition of e.m.f. and appreciate how (and why) it differs from that of potential difference

❑ know that the sum of the e.m.f.s is equal to the sum of the potential differences around any closed circuit loop and appreciate that this is a consequence of conservation of energy

❑ appreciate that all sources of e.m.f. will have some internal resistance

❑ realise that the terminal potential difference only equals the e.m.f. when no current flows

❑ appreciate the need for an infinite resistance voltmeter to measure e.m.f.

❑ understand why the terminal potential difference falls as more current is supplied by the power supply

❑ have learnt a description of an experiment to find the e.m.f. and internal resistance of a power supply

❑ appreciate the need for a car battery to have a very small internal resistance and for an e.h.t. supply to have a very large internal resistance

❑ can confidently solve circuit problems involving internal resistance

❑ are familiar with the 'general requirements' (see Appendix 1) and how they apply to the topic of electricity

Unit 2

Testing your knowledge and understanding

Quick test

Answers to these questions, together with explanations, are in the Answers section which follows Chapter 6.

Select the correct answer to each of the following questions from the four answers supplied. In each case only one of the four answers is correct. Allow about 40 minutes for the 20 questions.

1 Which of the following units is a measure of the rate of flow of charge?

    A ampere   **B** volt   **C** ohm   **D** watt

2 Four small conductors, on the edge of an insulating disk of radius r, are each given a charge q (Figure 2.14).
When the disk is rotated at n revolutions per second the equivalent electric current round the edge of the disk is

    A $4q/n$ **B** $8\pi rqn$ **C** $4qn$ **D** $2qn/\pi r$

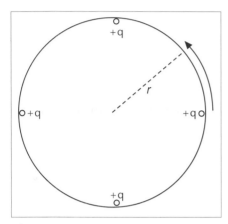

Fig 2.14

3 Which of the following statements concerning electric current is NOT correct?

    A The movement of electrons is a current
    B Current is directly proportional to the charge carrier density
    C Current is directly proportional to the drift velocity of the charge carriers
    D Current is inversely proportional to the charge on the carriers

4 When the potential difference across the ends of a copper wire is increased, the current also increases. Which of the following statements regarding n, the number of charge carriers per unit volume in the wire, and v, the drift velocity of the charge carriers, is correct?

    A Both n and v are increased
    B n is unaltered but v is increased
    C n is increased but v is unaltered
    D Both n and v are decreased

5 All the resistors in Figure 2.15 are identical and the voltmeters are ideal. The potential difference between points P and Q is 6 V in each case. Which voltmeter displays a reading of 3 V?

6 A potential difference of 15 V is applied across a resistor for a time interval of 5 s. The current in the resistor is 2 A. Which of the following statements is NOT correct?

    A The charge passed is 10 C
    B The potential difference is 15 J C^{-1}
    C The energy dissipated is 75 J
    D The resistance is 7.5 Ω

7 You are provided with three 10 Ω resistors. By combining **all three** resistors in various arrangements, which of the following resistances could NOT be obtained?

    A 3.3 Ω   **B** 6.7 Ω   **C** 15 Ω   **D** 20 Ω

8 Two pupils are asked to find the value of an unknown resistance. They each use the same apparatus – a d.c. supply of negligible internal resistance, an ammeter and a voltmeter. Both meters are NOT ideal. Figure 2.16 shows the circuits used – circuit J by Jane and circuit M by Margaret.
Jane obtained the results:
voltmeter reading = 5.0 V
ammeter reading = 1.0 mA

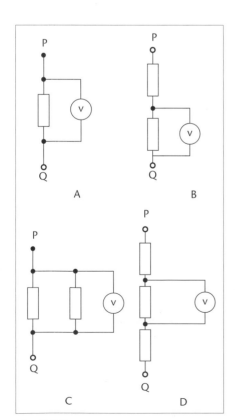

Fig 2.15

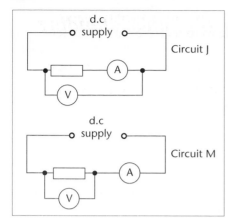

Fig 2.16

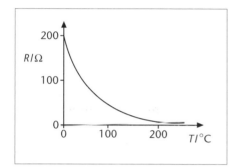

Fig 2.17

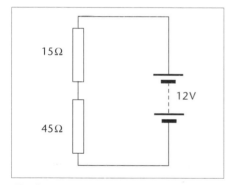

Fig 2.18

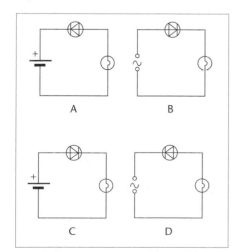

Fig 2.19

Margaret's results would be

   A voltmeter reading = slightly more than 5.0 V and ammeter reading = slightly more than 1.0 mA

   B voltmeter reading = slightly more than 5.0 V and ammeter reading = slightly less than 1.0 mA

   C voltmeter reading = slightly less than 5.0 V and ammeter reading = slightly more than 1.0 mA

   D voltmeter reading = slightly less than 5.0 V and ammeter reading = slightly less than 1.0 mA

9 The main reason for the increase in electrical resistance R of a metallic conductor when its temperature is raised is the

   A higher drift speed of the electrons

   B increase in the length of the conductor

   C reduction in the number of free electrons

   D increased amplitude of vibrations of lattice ions

10 The graph in Figure 2.17 shows the variation of the resistance R of a component.
The component could be a

   A light-dependent resistor

   B resistance wire

   C NTC thermistor

   D fuse

11 A wire of uniform cross-sectional area 0.5 mm^2 has a length of 10 m and a resistance of 3 Ω. The resistivity of the material from which it is made is

   A 1.5×10^{-7} Ω m

   B 1.5×10^{-4} Ω m

   C 6.7×10^{3} Ω m

   D 6.7×10^{6} Ω m

12 The table gives the resistance per unit length R of four wires of the same conducting material with different diameters d.

R/Ω m^{-1}	32	8.0	3.6	2.0
d/mm	0.5	1.0	1.5	2.0

Which of the following shows the relationship between R and d?

   A $R = kd$ **B** $R = kd^2$ **C** $R = k/d$ **D** $R = k/d^2$

13 Figure 2.18 shows a circuit in which a 15 Ω resistor and a 45 Ω resistor are connected to a 12 V battery.
Which of the following gives the potential difference across each resistor?

   A 9 V across 15 Ω and 3 V across 45 Ω

   B 3 V across 15 Ω and 9 V across 45 Ω

   C 6 V across 15 Ω and 6 V across 45 Ω

   D 0 V across 15 Ω and 12 V across 45 Ω

14 A 1.5 V lamp is connected with a diode in various arrangements to a 1.5 V a.c. or d.c. supply (Figure 2.19).
In which circuit does the lamp NOT light?

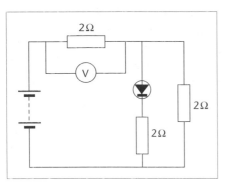

Fig 2.20

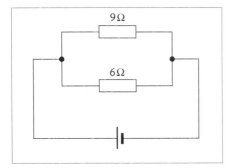

Fig 2.21

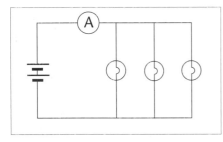

Fig 2.22

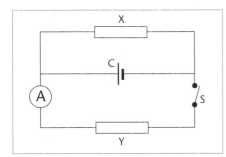

Fig 2.23

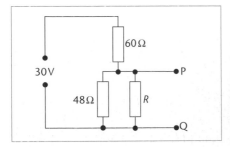

Fig 2.24

15 The diode in the circuit in Figure 2.20 has zero forward resistance. The voltmeter reads 12 V. If the diode were reversed the voltmeter reading would be

    A 6 V   **B** 8 V   **C** 9 V   **D** 18 V

16 The power developed in the 9 Ω resistor in Figure 2.21 is 36 W. The power developed in the 6 Ω resistor is

    A 18 W   **B** 24 W   **C** 48 W   **D** 54 W

17 The e.m.f. of a cell is the chemical energy transformed into electrical energy within the cell

    A   per ampere flowing in the cell
    B   per coulomb flowing through the cell
    C   per unit resistance in series with the cell
    D   when the cell is short-circuited

18 Figure 2.22 shows three identical lamps, which are being lit by a battery that has a significant internal resistance. The ammeter has negligible resistance.

If the filament of one of the lamps breaks, which of the following shows what happens to the ammeter reading and to the brightness of the remaining lamps?

    A   ammeter reading decreases and lamp brightness increases
    B   ammeter reading decreases and lamp brightness decreases
    C   ammeter reading increases and lamp brightness increases
    D   ammeter reading increases and lamp brightness decreases

19 Cell C in Figure 2.23 has an e.m.f. of 12 V and an internal resistance of 2 Ω.

X and Y are resistors, each of value 6 Ω. The ammeter A has negligible resistance. When switch S is closed, the reading of the ammeter is

    A 0.8 A   **B** 1.2 A   **C** 1.5 A   **D** 2.4 A

20 The potential divider circuit shown in Figure 2.24 is powered by a 30 V supply of negligible internal resistance.

The potential difference across P and Q is 5.0 V. The resistance of R is

    A 10 Ω   **B** 12 Ω   **C** 16 Ω   **D** 24 Ω

Worked examples

Study the following worked examples on electricity carefully. Make sure you fully understand their answers before attempting the practice assessment questions.

Worked example 1

A copper wire is 2.0 m long and has a cross-sectional area of 1.0 mm². It has a potential difference of 0.12 V across it when the current in it is 3.5 A. Draw a circuit diagram to show how you would check these voltage and current values.   **[3]**
Calculate the rate at which the power supply works on the wire.   **[3]**
Copper has about 1.7×10^{29} electrons per metre cubed. Calculate the drift velocity of the charge carriers in the wire.   **[2]**
The power from the supply connected to the wire is equal to the total force F_t on the electrons multiplied by the drift velocity at which the electrons travel. Calculate F_t.   **[2]**
                                      (Total 10 marks)
               (Edexcel Module Test PH1, January 1996, Q. 4)

Helpful hint

You must show a workable circuit that has an ammeter in series and a voltmeter in parallel with the copper wire.

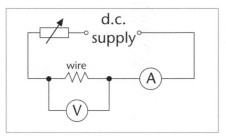

Fig 2.25

Answer:

Circuit diagram (Figure 2.25) showing   power supply + variable resistor ✓
    ammeter in series with wire ✓
    voltmeter in parallel with wire ✓

Rate of working on wire = power = IV ✓
    = 3.5 A × 0.12 V ✓
    = 0.42 W ✓

From $I = nAvq$
$v = I/nAq = 3.5\ \text{A}/(1.7 \times 10^{29}\ \text{m}^{-3} \times 1.0 \times 10^{-6}\ \text{m}^2 \times 1.6 \times 10^{-19}\ \text{C})$ ✓
$v = 1.3 \times 10^{-4}\ \text{m s}^{-1}$ ✓
Note the conversion of mm² to m².

Power = $F_t \times v$
F_t = power/v = 0.42 W/(1.3 × 10⁻⁴ m s⁻¹) ✓
    = 3260 N ✓

Worked example 2

Define the term *resistivity*.   **[1]**
The resistivity of copper is $1.7 \times 10^{-8}\ \Omega$ m. A copper wire is 0.60 m long and has a cross-sectional area of 1.0 mm². Calculate its resistance.   **[3]**
Two such wires are used to connect a lamp to a power supply of negligible internal resistance. The potential difference across the lamp is 12 V and its power is 36 W. Calculate the potential difference across each wire.   **[4]**
Draw a circuit diagram of the above arrangement. Label the potential difference across the wires, lamp and power supply.   **[3]**
(Total 11 marks)
(*Edexcel Module Test PH1, June 1996, Q. 5*)

Answer:
Since $R = \rho l/A$, $\rho = RA/l$
so   resistivity = resistance × cross-sectional area/length ✓

$R = \rho l/A$ ✓
    = $1.7 \times 10^{-8}\ \Omega$ m × 0.60 m/(1.0 × 10⁻⁶ m²) ✓
    = 0.0102 Ω ✓

Current in lamp (and wires) $I = P/V$ ✓
    = 36 W/(12 V) = 3 A ✓

Potential difference across each wire $V = IR$ ✓
    = 3 A × 0.0102 Ω = 0.0306 V ✓

Circuit diagram (Figure 2.26) showing   12 V across lamp ✓
    0.03 V across each wire ✓
    12.06 V across power supply ✓

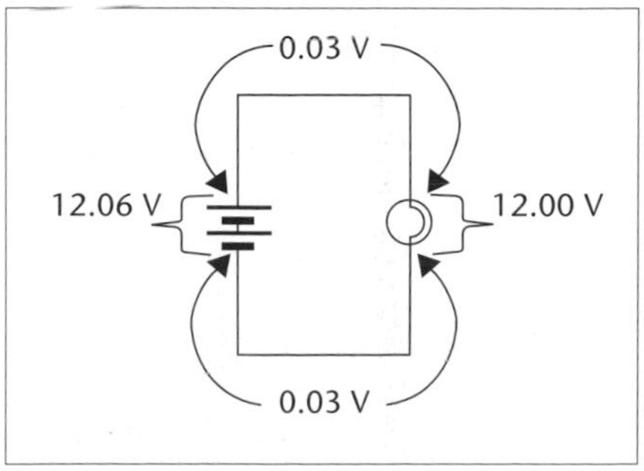

Note the use of the word equation definition.

Fig 2.26

Unit 2

Fig 2.27

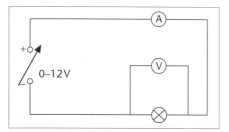

Fig 2.28

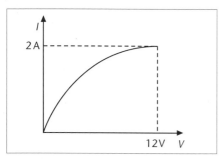

Fig 2.29

Answers to these questions, together with explanations, are in the Answers section which follows Chapter 6.

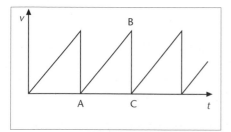

Fig 2.30

Worked example 3

You are asked to set up a circuit to take some measurements and to draw a graph which shows how the current in a 12 V, 24 W electric filament lamp varies with the potential difference across it.

Figure 2.27 shows the electrical components you will need. Complete a suitable circuit diagram by drawing the connection wires. **[2]**
What measurements would you make using this circuit? **[3]**
Sketch and label the graph you would expect to obtain. **[3]**
(Total 8 marks)
(*Edexcel Module Test PH1, January 2000, Q. 3*)

Answer:
Completed circuit diagram (Figure 2.28) showing
      ammeter in series with lamp and supply ✓
      voltmeter across lamp ✓

Record voltmeter reading ✓
and corresponding ammeter reading ✓
Repeat for a range of supply voltage settings *or* for a range of currents ✓

Sketch graph (Figure 2.29) showing   axes labelled *I* and *V* ✓
                          graph line with correct curvature ✓
                          12 V, 2 A correctly labelled ✓

Practice questions

The following are typical assessment questions on electricity. Attempt these questions under similar conditions to those in which you will sit your actual test.

1 The current *I* flowing through a conductor of cross-sectional area *A* is given by the formula $I = nAQv$ where *Q* is the charge on a charge carrier. Give the meanings of *n* and *v*. **[2]**
Show that the equation is homogeneous with respect to units. **[3]**
With reference to the equation, explain the difference between a metal conductor and a plastic insulator. **[2]**
(Total 7 marks)
(*Edexcel Unit Test PHY2, June 2001, Q. 1*)

2 A velocity–time graph for an electron is shown in Figure 2.30. Describe carefully the motion represented by lines AB and BC. **[2]**
The graph is a much-simplified model of how an electron moves along a wire carrying a current. Explain what causes the motion represented by AB and by BC. **[2]**
Explain the term *drift velocity* and indicate its value on the graph in Figure 2.30. **[2]**
With reference to the behaviour of the electron, explain why the wire gets warm. **[1]**
(Total 7 marks)
(*Edexcel Module Test PH1, June 2001, Q. 7*)

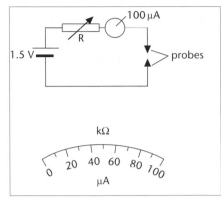

Fig 2.31

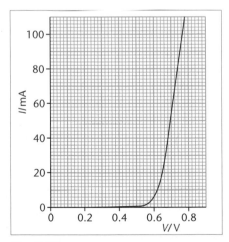

Fig 2.32

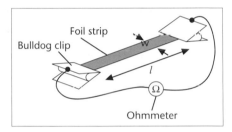

Fig 2.33

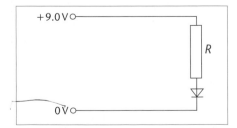

Fig 2.34

3 A 100 W tungsten filament lamp operates from the 230 V mains.
Calculate its resistance.   **[3]**
The drift speed of the electrons in the filament is much higher than
the drift speed of electrons in the rest of the circuit. Suggest and
explain a reason for this.   **[4]**
(Total 7 marks)
(Edexcel Module Test PH1, June 1999, Q. 5)

4 The diagram (Figure 2.31) shows a circuit for measuring resistance
(i.e. an ohmmeter).
Before any readings are taken, the two probes are connected together
and the variable resistor is adjusted so that the meter reads full-scale
deflection. Calculate the resistance of *R* for full-scale deflection.   **[3]**
With *R* fixed at this value, what *additional* resistance connected
between the probes would give a meter reading of (a) a half of full-
scale deflection and (b) a quarter of the full-scale deflection from the
0 μA end of the scale?   **[2]**
Use your calculated values to mark the scale in Figure 2.31 so that it
would show directly the resistance of any component connected
between the probe terminals. Mark the full range of the scale.   **[3]**
(Total 8 marks)
(Edexcel Module Test PH1, June 1997, Q. 4)

5 A student is planning an experiment to measure the resistivity of
aluminium. She plans to use an ohmmeter to measure the resistance
of a rectangular strip of aluminium foil fastened between two
bulldog clips (Figure 2.32).
She also intends to measure the thickness *t* of the foil and the length
l and width *w* of the strip.
Explain how she should calculate the resistivity from her
measurements.   **[2]**
The student decides that for sufficient accuracy the resistance of the
strip must be at least 1.0 Ω. To see what dimensions would be
suitable, she does some preliminary experiments using strips 20 mm
wide cut from foil 0.15 mm thick. She finds that for strips of a
convenient length the resistance is far too small.
Calculate the length of strip, 20 mm wide and 0.15 mm thick, which
would have a resistance of 1.0 Ω. (Resistivity of
aluminium = 2.7×10^{-8} Ω m)   **[3]**
Suggest a way, other than increasing its length, by which she could
increase the resistance of her strip. Comment on whether this
change would lead to a more precise measurement of the resistivity.
                                                                                               [2]
(Total 7 marks)
(Edexcel Unit Test PHY2, June 2001, Q. 2)

6 The graph (Figure 2.33) shows the current–voltage characteristic of a
semiconductor diode in forward bias.
State, with a reason, whether the diode obeys Ohm's law.   **[1]**
Show that the when the potential difference across the diode is
0.74 V its resistance is about 9 Ω.   **[2]**
When the diode is connected in the circuit shown (Figure 2.34), the
potential difference across it is 0.74 V.
Calculate the value of the resistance *R*.   **[3]**

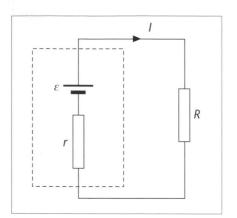

Fig 2.35

Electronic circuit designers often use a simple model of this type of diode. This 'model diode' has the following properties:

(i)   for any voltage below +0.7 V it does not conduct at all.

(ii) once the voltage reaches +0.7 V the diode can pass any size of current with no further increase in the voltage.

Add a second graph to the grid in Figure 2.33 to show the current–voltage characteristic of this model diode.   **[2]**

(Total 8 marks)

(Edexcel Unit Test PHY2, June 2001, Q. 4)

7 The diagram (Figure 2.35) shows a cell, of e.m.f. ε and internal resistance r, driving a current I through an external resistance R. Using these symbols, write down a formula for

(i)   the power dissipated in the external resistance

(ii) the power dissipated in the internal resistance

(iii) the rate of conversion of chemical energy in the cell   **[3]**

Using these formulae, write down an equation expressing conservation of energy in the circuit, and hence show that $I = \varepsilon/(R + r)$.   **[2]**

The equation $I = \varepsilon/(R + r)$ shows that the internal resistance of a power supply limits the current which can be drawn from it. Explain this.   **[2]**

A 5 kV laboratory supply can be made safe for student use by connecting an internal series resistor. The following resistors are available:

1 kΩ   10 kΩ   100 kΩ   1 MΩ

Explain which resistor should be used to make the supply as safe as possible.   **[2]**

(Total 9 marks)

(Edexcel Unit Test PHY2, June 2001, Q. 3)

8 A torch has three identical cells, each of e.m.f. 1.5 V, and a lamp that is labelled 3.5 V, 0.3 A. Draw a circuit diagram for the torch.   **[2]** Assume that the lamp is lit to normal brightness and that the connections have negligible resistance. Mark on your diagram the potential difference across each circuit component and the current flowing in the lamp.   **[3]**

Calculate the internal resistance of one of the three cells.   **[3]**

(Total 8 marks)

(Edexcel Module Test PH1, January 2000, Q. 5)

Part 2 Thermal physics

Introduction

Thermal physics involves the transfer of energy, either by heating or working, together with an understanding of the effects that this produces. As you study thermal physics, you learn how the motion of the molecules of a gas is responsible for the pressure it exerts. You find that the speed of the molecules increases with temperature and produces an increased pressure. The resulting relationship allows you to predict the temperature at which the pressure of any gas would become zero. You also discover how the pressure and volume of a gas relate when at a constant temperature. You find that transferring energy to a solid either increases its temperature or changes it to a liquid and so appreciate the difference between specific heat capacity and specific latent heat. You learn about internal energy and how conservation of energy leads to the first law of thermodynamics, and discover the differences between heating, electrical working and mechanical working. You find that energy flowing naturally from hot to cold can do work while work must be done to force energy to flow the other way and learn how to improve the maximum efficiency of such systems.

Things to understand

Pressure and temperature

- solids are rigid and can be used to transmit forces (Figure 2.36a)
- fluids (gases and liquids) flow to fit the shape of their container and can be used to transmit pressure (Figure 2.36b)
- pressure acts at 90° to any surface and is the force exerted per unit area
- the unit of pressure is the pascal (Pa)
- a calibrated thermometer is used to measure temperature
- the pressure of a fixed amount of gas in a given volume increases with temperature
- reducing temperature, reduces pressure and experimental work predicts that all gases have the same temperature (absolute zero) at which their pressures would be zero
- pressure is directly proportional to Kelvin temperature, $p \propto T$
- the Kelvin and Celsius temperature scales are related by $T/\text{K} = \theta/°\text{C} + 273$

The ideal gas equation

- the pressure of a fixed amount of gas at constant temperature increases as its volume decreases
- experimental work shows that pressure and volume are inversely proportional, $p \propto 1/V$ if T is constant

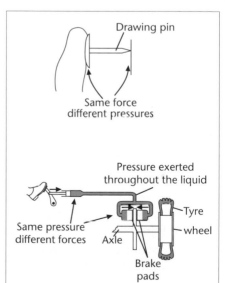

Fig 2.36 (a) The solid drawing pin transmits the force, (b) The fluid in the brake pipes transmits pressure

- if all three macroscopic properties change, $p \propto T/V$ so that pV/T = constant that depends only on the number of moles of gas present
- for one mole, $pV/T = R$, the molar gas constant
- for n moles, $pV = nRT$

Kinetic model of a gas

- Brownian motion gives evidence for the random motion of air particles
- high compressibility of gases gives evidence for the large spacing of air molecules
- the particles of all gases are moving around continually at high speed
- pressure results from collisions of the gas particles with the container walls
- the speed of gas particles increases with temperature
- pressure increases with temperature as the collisions with the walls are both harder and more frequent
- a decrease in volume results in an increase in the packing density of particles so there are more collisions per unit area in the same time and pressure therefore rises
- the behaviour of a gas can be modelled using a mechanical model (Figure 2.37)
- a theoretical model can be used to predict the pressure that a gas will exert on its container
- the theoretical model relies on a number of assumptions:

    gases consist of identical molecules in continuous random motion
    molecular collisions are, on average, elastic
    the volume of the actual molecules is negligible compared to the volume of the container
    the molecules only exert a force on each other during collisions

- the theoretical prediction only agrees with the ideal gas equation if the average molecular kinetic energy ($\frac{1}{2}m<c^2>$) is proportional to the Kelvin temperature
- the molecules of a gas have a wide range of speeds

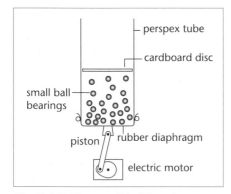

Fig 2.37 *The small ball bearings represent gas molecules*

Labels in figure: perspex tube, cardboard disc, small ball bearings, piston, rubber diaphragm, electric motor

Heating solids and liquids

- internal energy is the sum of the molecular kinetic and potential energies
- adding energy to a substance increases its internal energy
- increasing the internal energy of a body either raises its temperature or changes its state, from solid to liquid or from liquid to gas
- the energy needed to raise the temperature of an object is proportional to both the mass of the object and the temperature rise and depends on the material from which it is made
- specific heat capacity is measured in $J\,kg^{-1}\,K^{-1}$
- the energy needed to change the state of an object is proportional to the mass of the object and depends on the material from which it is made
- specific latent heat is measured in $J\,kg^{-1}$
- more energy is needed to change the state of an object than to increase its temperature by 1 °C

- changing from a liquid to a gas requires much more energy than changing from a solid to a liquid as the molecules need to be pulled apart against intermolecular attractive forces and the substance has to do more work pushing back the surroundings as it expands

Heating and working

- a hot object has a greater concentration of internal energy than a cold object
- when two bodies are in thermal contact there is a random exchange of energy between them
- this random exchange of energy always results in an energy flow ΔQ from hot to cold
- the process where energy flows through a temperature difference from hot to cold is called heating
- energy transfer can occur through conduction, convection, radiation and evaporation
- working is the process where a force causes motion

  mechanical working is where a mechanical force causes motion of a mass, $\Delta W = F\Delta x$
  electrical working is where an electrical force causes motion of a charge, $\Delta W = V\Delta Q$

- working can transfer energy ΔW to any object, hot or cold, while heating can only transfer energy from a hot object to one that is colder
- since energy is conserved, change in internal energy of a body equals the sum of the energy transferred by heating and the energy transferred by working
- heating and working are positive where the process adds energy to the body and negative where the process transfers energy away from the body
- energy transferred by heating is zero in either an isolated system or where there is no temperature difference, $\Delta Q = 0$
- there is no change in internal energy, $\Delta U = 0$, of a body if its temperature is constant and it is not changing state

Heat engines and pumps

- a heat engine takes energy from a hot source and uses some of it to do work
- the rest of the energy is transferred by heating to a cold sink
- efficiency is the ratio of the useful work output to the total energy taken from the hot source (Figure 2.38)
- maximum thermal efficiency depends on the Kelvin temperatures T_1 of the hot source and T_2 of the cold sink
- efficiency is increased either by increasing the temperature of the hot source or by reducing the temperature of the cold sink
- although energy flows naturally from hot to cold, a heat pump does work to move energy from cold to hot
- refrigerators, freezers and air conditioners all use heat pumps
- the energy expelled to the hot surroundings is the sum of the energy removed from the cold body and the work needed to remove it
- to maintain a cold body at a constant temperature, a heat pump has to remove energy from the body at the same rate as energy is entering it

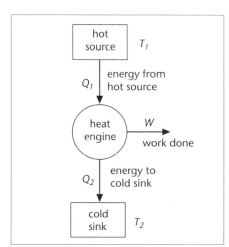

Fig 2.38 Maximum efficiency = $W/Q_1 = 1 - (T_2/T_1)$

Things to learn

You should learn the following for your Unit PHY2 Test. Remember that it may also test your understanding of the 'general requirements' (see Appendix 1).

Equations that will *not* be given to you in the test

❑ pressure = force/area

$$p = F/A$$

❑ pressure × volume = number of moles × molar gas constant × absolute temperature

$$pV = nRT$$

❑ pressure × volume/temperature = constant

$$p_1 V_1/T_1 = p_2 V_2/T_2$$

Laws

❑ pressure law: for a fixed mass of gas at constant volume, the pressure is directly proportional to the Kelvin temperature

❑ Boyle's law: for a fixed mass of gas at constant temperature, pressure × volume is constant

❑ first law of thermodynamics: increase in internal energy (ΔU) = energy transferred by heating (ΔQ) + energy transferred by working (ΔW)

General definitions

❑ absolute zero: temperature at which the pressures of all gases would be zero (since kinetic energy of molecules = 0); the lowest temperature theoretically possible

❑ Brownian motion: random motion of visible particles caused by random impacts from invisible molecules

❑ mean square speed $<c^2>$: sum of the squares of the individual molecular speeds divided by the total number of molecules

❑ r.m.s. (root mean square) speed: square root of the mean square speed

❑ internal energy: sum of the random kinetic and potential energies of the molecules of a body

❑ specific heat capacity: energy needed to raise the temperature of 1 kg of that substance by 1 K without a change of state

❑ specific latent heat of fusion: energy needed to change 1 kg of solid into liquid at its melting point

❑ specific latent heat of vaporisation: energy needed to change 1 kg of liquid into vapour at its boiling point

❑ heating: process in which energy transfer is driven by a temperature difference with the energy flowing from hot to cold

❑ mechanical working: process in which energy transfer occurs when a force moves through a distance

❑ electrical working: process in which energy transfer occurs when a force moves a charge

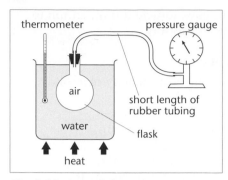

Fig 2.39 *Record the pressure at different temperatures*

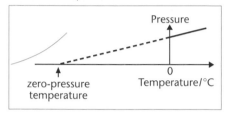

Fig 2.40 *Pressure would be zero at −273 °C*

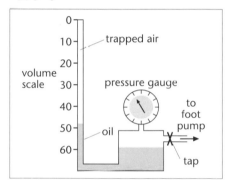

Fig 2.41 *Record the pressure at different volumes*

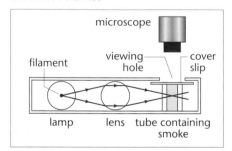

Fig 2.42 *The smoke particles have a random, zig-zag motion*

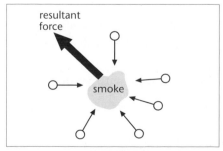

Fig 2.43 *The resultant force changes direction from one moment to the next*

- ❑ heat engine: a device that takes energy from a hot source, uses some of this to do mechanical work, and gives the rest to a cold sink
- ❑ heat pump: a device that does work to move internal energy from a cold body to a hot body

Word equation definitions

- ❑ pressure = force/area
- ❑ specific heat capacity = energy supplied/(mass × temperature rise)
- ❑ specific latent heat = energy supplied/mass of substance that has changed state
- ❑ efficiency = useful energy output/total energy input

Experiments

❑ 1. The pressure law

Submerge as much of the flask as possible in the water (Figure 2.39).
Use a short length of tubing to connect to pressure gauge since the air in it will not get fully heated.
Record a series of corresponding readings of pressure and temperature for temperatures from 0 to 100 °C.
Allow time for the air in the flask to reach the same temperature as the water.
Plot a graph of pressure against temperature and use it to predict the temperature at which the pressure would become zero (Figure 2.40).
Using absolute zero as origin, the graph shows that pressure is directly proportional to the Kelvin temperature.

❑ 2. Boyle's law

Measure the pressure and the volume of the trapped air (Figure 2.41).
Use a foot pump to increase the pressure.
Record a series of corresponding readings of pressure and volume.
Allow time between readings for the compressed air to return to room temperature.
Multiply each pressure by its corresponding volume.
Results show that 'pressure × volume = constant'.
Alternatively, plot a graph of pressure against 1/volume.
Straight-line graph through the origin shows that pressure is inversely proportional volume.

❑ 3. Brownian motion

Use a cover slip to trap smoke from a burning straw in the smoke cell (Figure 2.42).
The lens focuses light from the lamp onto the smoke particles.
Light reflects into the microscope from the smoke particles that appear as very small bright dots.
Smoke particles dance about randomly, moving first one way and then immediately another.
The much smaller air molecules are knocking the smoke particles about.
Due to its small size, there is an imbalance in the distribution of the air molecules hitting the smoke particle at any instant (Figure 2.43).
The constantly changing resultant force moves the smoke particle first one way and then another.

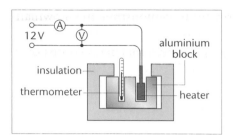

Fig 2.44 *The electric heater supplies energy to the object*

Note that the equation has been rearranged to give the required quantity and that all symbols used have been previously defined

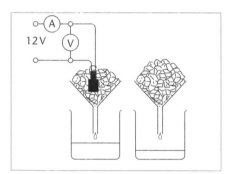

Fig 2.45 *The electric heater supplies additional energy to the ice*

Air molecules must be moving very fast to have sufficient momentum to cause the heavier smoke particles to move in this way.

☐ 4. Measuring specific heat capacity

A suitable apparatus for measuring the specific heat capacity of a solid object is shown in Figure 2.44.

Measure the mass m of the object and its initial temperature θ_1.

Switch on the electric heater for a measured time t, e.g. 5 min.

Measure the potential difference V and the current I.

The electrical work done on the heater is calculated using VIt.

Since the object is well lagged, it is assumed that all this energy is transferred to the object.

Measure the highest temperature θ_2 reached after the heater is switched off.

Since energy supplied = mass × specific heat capacity × temperature rise

$$\text{specific heat capacity} = VIt/[m \times (\theta_2 - \theta_1)$$

☐ 5. Measuring specific latent heat

A suitable apparatus for measuring the latent heat of fusion of water is shown in Figure 2.45.

Switch on the electric heater and measure the potential difference V and the current I.

Measure the masses of two empty beakers.

Use the beakers to collect water from each funnel for a measured time t e.g. 5 min.

Calculate the mass of water in each beaker and so find the additional mass m of ice melted by the heater.

The electrical work done on the heater is calculated using VIt.

Assuming all energy from the heater is used to melt the ice and since energy supplied = mass × specific latent heat

$$\text{specific latent heat} = VIt/m$$

✳ *Checklist*

Before attempting the following questions on thermal physics, check that you:

- ☐ appreciate that solids transmit forces and fluids transmit pressures
- ☐ know that pressure is the force per unit area and is measured in pascals (Pa) where $1\ \text{Pa} = 1\ \text{N m}^{-2}$
- ☐ have learnt a description of an experiment showing how the pressure of a gas varies with temperature
- ☐ can sketch a graph showing the variation of pressure with temperature
- ☐ have learnt a statement of the pressure law
- ☐ understand the concept of absolute zero
- ☐ know how to convert between the Kelvin and Celsius temperature scales
- ☐ have learnt a description of an experiment showing how the pressure of a gas varies with volume
- ☐ can sketch a graph showing the variation of pressure with volume of a gas at constant temperature
- ☐ have learnt a statement of Boyle's law
- ☐ can solve problems using the ideal gas equation, $pV = nRT$

❏ have learnt a description of an experiment to demonstrate the Brownian motion of smoke particles in air

❏ can explain how Brownian motion gives evidence that gases consist of atoms or molecules moving randomly at high speeds

❏ know that molecular speed increases with temperature

❏ appreciate that pressure results from collisions of the gas particles with the walls of their container

❏ can explain in molecular terms why the pressure of a gas increases with increasing temperature and with decreasing volume

❏ have learnt the assumptions on which the theoretical model of a gas is founded

❏ am familiar with the steps involved in using the theoretical model to derive the equation $p = \frac{1}{3}\rho <c^2>$

❏ can use this equation to calculate the root mean square speed of gas molecules

❏ understand the difference between mean speed and r.m.s. speed

❏ know that the theoretical equation agrees with the ideal gas equation provided that the average molecular kinetic energy is directly proportional to the Kelvin temperature

❏ appreciate that internal energy is the sum of the molecular kinetic and potential energies

❏ know that either a rise in temperature or a change in state increases internal energy

❏ have learnt the definition of specific heat capacity

❏ have learnt a description of an experiment using an electric heater to measure the specific heat capacity of a solid and can adapt this to measure that of a liquid

❏ can identify sources of experimental error in such heating experiments and know some ways of reducing these

❏ can calculate the amount of energy transferred when a body either warms up or cools down

❏ have learnt the definition of specific latent heat

❏ have learnt a description of an experiment using an electric heater to measure the specific latent heat of fusion of water and can adapt this to measure its latent heat of vaporisation

❏ can calculate the amount of energy transferred when a body changes state

❏ appreciate why, for the same substance, the latent heat of vaporisation is much greater than the latent heat of fusion

❏ know that the random exchange of energy between a hot and cold body results in a net energy flow from hot to cold, a process known as heating

❏ can explain how energy transfer occurs by conduction, convection, radiation and evaporation

❏ appreciate that working involves a force causing motion and that working can be either mechanical or electrical

❏ understand the difference between heating and working

❏ have learnt a statement of the first law of thermodynamics

❏ appreciate that this law is simply an application of the principle of conservation of energy

❑ can apply this law in the form $\Delta U = \Delta Q + \Delta W$ to any given system

❑ appreciate the conditions under which ΔU and ΔQ are zero and the significance of positive and negative values of the quantities involved

❑ understand the difference between a heat engine and a heat pump

❑ can calculate the maximum thermal efficiency of a heat engine and know how to maximise its value

❑ appreciate the part played by heat pumps in the operation of refrigerators, freezers and air conditioners

❑ understand that to maintain a constant temperature, a heat pump must remove energy from a system at the same rate as it is entering

❑ are familiar with the 'general requirements' (see Appendix 1) and how they apply to the topic of thermal physics

 Testing your knowledge and understanding

Quick test

Answers to these questions, together with explanations, are in the Answers section which follows Chapter 6.

Select the correct answer to each of the following questions from the four answers supplied. In each case only one of the four answers is correct. Allow about 40 minutes for the 20 questions.

1 Which of the following is NOT a possible unit of pressure?

  A $N\,m^{-2}$ **B** Pa **C** $kg\,m^{-3}$ **D** $J\,m^{-3}$

2 A wooden block of mass 50 kg rests on a table. The area of the base of the block is $0.5\ m^2$. The pressure the block exerts on the table is about

  A 25 Pa **B** 100 Pa **C** 250 Pa **D** 1000 Pa

3 The pressure of a fixed mass of gas is 210 kPa at a temperature of 77 °C. Its pressure at a temperature of 27 °C is

  A 74 kPa **B** 180 kPa **C** 245 kPa **D** 600 kPa

4 Which of the following mathematical relationships is a correct representation of Boyle's law?

  A $pV = $ constant
  B $pT = $ constant
  C $p/V = $ constant
  D $p/T = $ constant

5 It is observed that the volume of an air bubble increases as it rises to the surface of a lake. The most reasonable explanation of this observation is that

  A the bubble takes in more air as it rises
  B the water is denser at the surface than at the bottom of the lake
  C the temperature of the air bubble decreases as it rises
  D the pressure on the air bubble is less at the surface than at the bottom of the lake

6 The ideal gas equation may be written $pV = nRT$ where the symbols have their usual meanings. The symbol n represents

  A the number of molecules in a mole
  B the number of molecules per unit volume
  C the number of moles present
  D the total number of molecules present

7 Which of the following is NOT a valid assumption of the kinetic model of an ideal gas?

    A The volume of the molecules is negligible compared with the volume of the gas

    B The molecules suffer negligible change of momentum on collision with the walls of the container

    C The molecular diameter is small compared with the average distance between molecules

    D Forces between molecules are negligible except during collisions

8 The following equations apply to an ideal gas where each symbol has its usual meaning.

$$pV = nRT \qquad p = \tfrac{1}{3}\rho <c^2> \qquad \tfrac{1}{2}m<c^2> = \tfrac{3}{2}kT$$

Which of the following is correct?

    A $<c^2>$ is the square of the mean speed of the molecules

    B k has no units

    C R has the units $J\,K^{-1}\,mol^{-1}$

    D ρ is independent of the number of gas molecules present

9 At the surface of the Earth, the density of air is $1.2\ kg\ m^{-3}$ and the atmospheric pressure is 1.0×10^5 Pa. The root mean square speed of the air molecules is

    A $500\ m\ s^{-1}$

    B $600\ m\ s^{-1}$

    C $250\,000\ m\ s^{-1}$

    D $360\,000\ m\ s^{-1}$

10 The speeds of five molecules in $km\ s^{-1}$ are 1, 2, 3, 4 and 5. The r.m.s. speed of the molecules is

    A $3\ km\ s^{-1}$ **B** $\sqrt{11}\ km\ s^{-1}$ **C** $\sqrt{55}\ km\ s^{-1}$ **D** $9\ km\ s^{-1}$

11 The curve in Figure 2.46 shows the distribution of speeds of a fixed number of molecules in a gas.
If the temperature of the gas is reduced, the peak of the curve

    A rises and shifts to the left

    B rises and shifts to the right

    C falls and shifts to the left

    D falls and shifts to the right

12 The internal energy of a fixed mass of an ideal gas is a function of

    A the pressure and the temperature, but not of the volume

    B the temperature and the volume, but not of the pressure

    C the pressure, but not of the temperature or the volume

    D the temperature, but not of the pressure or the volume

13 Which of the following, all initially at a temperature of 20 °C, will have absorbed the greatest quantity of energy?

    A A steel needle heated to red heat

    B A kettle of water heated to boiling point

    C A bath of water heated to 50 °C

    D A piece of lead shot heated to its melting point

14 The specific heat capacity of water is $4200\ J\ kg^{-1}\ K^{-1}$. How long will it take a 1 kW immersion heater to raise the temperature of 2 kg of water from 20 to 100 °C? (Assume that the energy supplied is all used in heating the water.)

    A 5.6 min   **B** 11.2 min   **C** 14.0 min   **D** 672 min

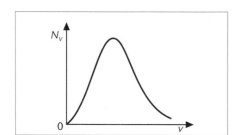

Fig 2.46

15 Two identical lead masses collide with each other when travelling in opposite directions at equal speeds of 300 m s^{-1}. The two masses join together and come to rest. If all the kinetic energy is used to heat the lead, what is the rise in temperature of the lead? (Assume that the specific heat capacity of lead is 500 J kg^{-1} K^{-1}.)

    A 9 °C   **B** 45 °C   **C** 90 °C   **D** 180 °C

16 In an experiment to measure the specific heat capacity of a liquid, the liquid is placed in a beaker and energy is supplied to it electrically using an immersion heater. The value obtained is found to be less than the accepted value. A reason for this could be that

    A no allowance was made for energy losses to the surroundings from the beaker

    B the ammeter measuring the current in the heating coil was faulty and read too high

    C no allowance was made for the energy used to warm up the beaker and the thermometer

    D the voltmeter measuring the potential difference across the heating coil was faulty and read too low

17 The graph (Figure 2.47) shows how the temperature of 1 kg of a substance, which is initially a solid, varies with time as the substance is heated uniformly at a rate of 2000 J min^{-1}. There are no heat losses. Which of the following is correct?

    A The specific latent heat of fusion of the substance is 6000 J kg^{-1}

    B The substance has fully changed to liquid after 4 min of heating

    C The specific heat capacity of the substance is greater when liquid than when solid

    D The substance must contain impurities

18 A 60 W electric light has been switched on for some time. Which of the following is NOT true for the next 30 min of its operation?

    A total internal energy of the lamp remains constant

    B electrical work done on the lamp is 1800 J

    C energy transferred to the lamp by heating is negative

    D change in internal energy = energy transferred by heating + energy transferred by working

19 In a power station, the temperature of the steam entering the turbine is 550 °C whereas that of the steam leaving is 180 °C. Assuming that it behaves as a heat engine, its maximum thermal efficiency is about

    A 33%   **B** 45%   **C** 55%   **D** 67%

20 On a hot day, energy enters through the walls of a refrigerator at a rate of 30 W. In order to maintain a constant internal temperature, the rate at which energy is released back into the kitchen by the heat pump is

    A zero

    B less than 30 W

    C exactly 30 W

    D more than 30 W

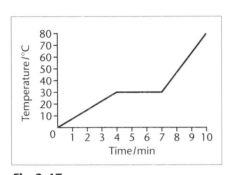

Fig 2.47

Worked examples

Study the following worked examples on thermal physics carefully. Make sure you fully understand their answers before attempting the practice assessment questions.

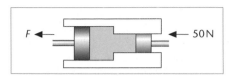

Fig 2.48

Worked example 1

Figure 2.48 shows two linked cylinders with internal diameters of 12 and 5 mm respectively.
A force of 50 N is applied to the piston in the narrow cylinder. The space between the pistons is filled with water. Calculate the magnitude of the force F exerted by the larger piston. [3]
On a very cold day the water freezes. State and explain the magnitude of the force F on such a day. [2]
(Total 5 marks)
(*Edexcel Module PH2, 1994 Specimen, Q. 4 (2nd part amended)*)

Answer:
Liquids transmit pressures
Pressure = force/area ✓
$p = 50 \text{ N}/[\pi \times (2.5 \times 10^{-3} \text{ m})^2] = 2.55 \times 10^6 \text{ N m}^{-2}$ ✓
$F = pA = 2.55 \times 10^6 \text{ N m}^{-2} \times \pi \times (6 \times 10^{-3} \text{ m})^2 = 288 \text{ N}$ ✓

$F = 50 \text{ N}$ ✓
since solids transmit forces ✓

Worked example 2

According to kinetic theory, the pressure p of an ideal gas is given by the equation
$$p = \tfrac{1}{3}\rho<c^2>$$
where ρ is the gas density and $<c^2>$ is the mean squared speed of the molecules.
Express ρ in terms of the number of molecules N, each of mass m, in a volume V. [1]

It is assumed in kinetic theory that the mean kinetic energy of a molecule is proportional to kelvin temperature T. Use this assumption, and the equation above, to show that under certain conditions p is proportional to T. [2]
State the conditions under which p is proportional to T. [2]
A bottle of gas has a pressure of 303 kPa above atmospheric pressure at a temperature of 0 °C. The bottle is left outside on a very sunny day and the temperature rises to 35 °C. Given that atmospheric pressure is 101 kPa, calculate the new pressure of the gas inside the bottle. [3]
(Total 8 marks)
(*Edexcel Unit Test PHY2, June 2001, Q. 6*)

Answer:
ρ = total mass/volume = Nm/V ✓

You need to link the given equation to kinetic energy ($\tfrac{1}{2} mv^2$)
e.g. $p = \tfrac{1}{3}\rho<c^2> = \tfrac{1}{3} Nm<c^2>/V = (\tfrac{2}{3}) \tfrac{1}{2} m<c^2> N/V$ ✓
so $p \propto \tfrac{1}{2} m<c^2> \propto T$ ✓
Even if you can't answer the previous part, you can still do this provided you have learnt a statement of the pressure law
For a fixed mass of gas (constant N) ✓
at constant volume (constant V) ✓

$p_1 = (303 + 101)$ kPa $T_1 = (0 + 273)$ K $T_2 = (35 + 273)$ K ✓
using $p_1/T_1 = p_2/T_2$
$p_2 = p_1 T_2/T_1 = 404$ kPa $\times 308$ K/(273 K) ✓
$p_2 = 456$ kPa ✓

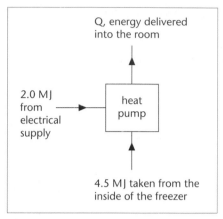

Fig 2.49

Worked example 3

The working part of a freezer is a heat pump, which pumps energy from the inside of a freezer to the outside. Figure 2.49 shows the energy flows for one day of operation.
What is the value of Q, the energy flow out of the freezer?   **[1]**
State the physical law you used to calculate your answer.   **[1]**
Suggest a reason why you need an energy source to pump energy from the inside of the freezer to the outside.   **[1]**
What is the power flow through the walls of the freezer?   **[3]**
Four containers of liquid milk, each having a mass of 2.3 kg and initially at 0 °C, are placed in the freezer. The specific latent heat of fusion of milk is 334 kJ kg^{-1}.
Calculate the additional energy that the heat pump must remove from the freezer as the milk freezes.   **[2]**
Inside the freezer there are no cooling fins at the bottom, but there are a large number of them towards the top. Explain how these fins cool the freezer and why there are none at the bottom.   **[2]**
(Total 10 marks)
(*Edexcel Module Test PH3, January 1997, Q. 5 (most*))

Answer:
$Q = (2.0 + 4.5)$ MJ = 6.5 MJ   ✓
Law of conservation of energy   ✓

Energy has to be forced to flow up a temperature gradient, from cold to hot   ✓

Power flow = energy flow/time   ✓
= 4.5×10^6 W/[$(24 \times 60 \times 60)$ s]   ✓
= 52 W   ✓

Energy removed = $l \Delta m = 334 \times 10^3$ J kg$^{-1} \times (4 \times 2.3$ kg)   ✓
= 3.1×10^6 J   ✓

Energy transfers from the hot air to the colder fins   ✓
due to convection the warm air will be at the top and cold air at bottom   ✓
fins are placed in the warm air at the top   ✓   **Max 2**

Answers to these questions, together with explanations, are in the Answers section which follows Chapter 6.

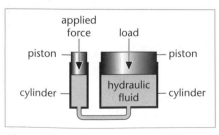

Fig 2.50

Practice questions

The following are typical assessment questions on thermal physics. Attempt these questions under similar conditions to those in which you will sit your actual test.

1   Figure 2.50 shows a simplified hydraulic jack.
With reference to the diagram, explain the principle of how a hydraulic jack may be used to raise the chassis of a car. You may be awarded a mark for the clarity of your answer.   **[4]**
(Total 4 marks)
(*Edexcel Unit Test PHY2, June 2001, Q. 8*)

2   A quantity of air is contained in a gas-tight syringe. The piston is clamped so that the volume of the air is fixed at 50 cm^3. When the air is at 0 °C its pressure is 1.00×10^5 Pa. The apparatus is now heated to 100 °C. Calculate the pressure of the air at 100 °C.   **[3]**
Draw an accurate graph to show how the air pressure varies with temperature over the range 0 to 100 °C. Label this graph A.   **[2]**

The piston is pushed in until the air volume is 25 cm^3. The piston is then clamped. On the same axes draw a second graph, labelled B, to show how the pressure would now vary over the same temperature range. **[2]**

(Total 7 marks)

(Edexcel Module Test PH3, January 2001, Q. 3)

3 Sketch the pressure-volume graph for a fixed mass of air at room temperature. **[2]**

In terms of the behaviour of molecules, explain qualitatively the shape of your graph. **[2]**

Figure 2.51 shows apparatus that could be used to check the shape of your graph.

How would you calculate the pressure of the air in the syringe? **[3]** Suggest one possible source of error in this experiment, other than errors in scale readings. **[1]**

(Total 8 marks)

(Edexcel Module Test PH3, June 2001, Q. 3)

4 The kinetic theory of gases is based on a number of assumptions. One assumption is that the average distance between the molecules is much larger than the molecular diameter. A second assumption is that the molecules are in continuous random motion. State and explain one observation in support of each assumption. **[2,2]**

(Total 4 marks)

(Edexcel Module Test PH3, January 1999, Q. 4)

5 A thin beaker is filled with 400 g of water at 0 °C and placed on a table in a warm room. A second identical beaker, filled with 400 g of an ice-water mixture, is placed on the same table at the same time. The contents of both beakers are stirred continuously. Figure 2.52 shows how the temperature of the water in the *first* beaker increases with time.

Use the graph to find the initial rate of rise of water temperature. **[2]** The specific heat capacity of water is 4200 J kg^{-1} K^{-1}. Use your value for the rate of rise of temperature to estimate the initial rate at which this beaker of water is taking in heat from the surroundings. **[3]** Figure 2.53 shows the temperature of the water in the *second* beaker from the moment it is placed on the table.

How do you explain the delay of 27 min before the ice-water mixture starts to warm up? **[2]** The specific latent heat of fusion of ice is 330 kJ kg^{-1}. Estimate the mass of ice initially present in the ice-water mixture. **[4]**

(Total 11 marks)

(Edexcel Module Test PH3, January 2000, Q. 2)

6 A well-insulated vessel contains 0.20 kg of ice at −10 °C. The graph in Figure 2.54 shows how the temperature of the ice would change with time if it were heated at a steady rate of 30 W and the contents were in thermal equilibrium at every stage.

Describe in terms of molecules the change which occurs between points P and Q. **[2]** Use the graph to determine the specific latent heat of fusion of water. **[3]**

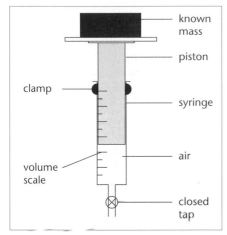

Fig 2.51

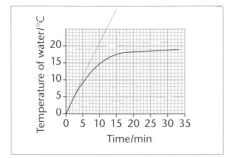

Fig 2.52

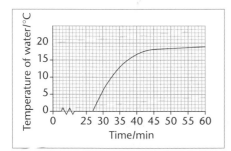

Fig 2.53

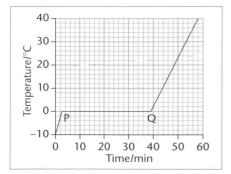

Fig 2.54

A student tries to plot this graph experimentally. He places crushed ice at –10 °C in a well-insulated beaker containing a small electric heater. What additional equipment would he need, and how should he use it, to obtain the data for his graph? **[2]**

Suggest one precaution he should take to try to get an accurate graph. **[1]**

Gallium is a metal with a melting point of 29 °C. Its specific heat capacity, in both the solid and liquid states, and its specific latent heat of fusion, are all smaller than those of water. Add to the graph in Figure 2.54 a second line showing the results you would expect if 0.2 kg of gallium, initially at –10 °C, was heated at the same rate of 30 W. **[3]**

(Total 11 marks)

(Edexcel Unit Test PHY2, June 2001, Q. 5)

7 A 60 W tungsten filament bulb is operating at its normal working temperature of 1600 °C. The equation $\Delta U = \Delta Q + \Delta W$ may be applied to the lamp filament. State and explain the **value** of each quantity for a 10 s period of operation. **[2,2,2]**

(Total 6 marks)

(Edexcel Unit Test PHY2, June 2001, Q. 7)

8 What is meant by a *heat engine*? **[3]**

Explain why there is a constant search for materials to make turbine blades that will operate at higher temperatures. **[1]**

(Total 4 marks)

(Edexcel Module Test PH3, January 1997, Q. 4)

3 Topics and the AS practical test

Part ❶ Topics

✳ Introduction

Unit 3 of the specification contains four different topics, although you have to study only one of them. The topic test paper contains four long structured questions together with spaces for your answers. Each question usually takes up four sides, consists of about five separate parts and is worth a total of 32 marks. You only have to answer one of these four questions so much of your completed question-answer booklet will remain blank! Each question mainly tests knowledge and understanding of the material in that topic and how this may be applied to other situations, although there may be a few marks for work based on material from Units 1 and 2. You are expected to be able to recall the meanings of specialist words and phrases from your chosen topic. Each question may involve quantitative work using supplied data and interpretation of supplied information, and will usually require part of the answer to be given in free prose for which there will be a quality of written communication mark. About 30% of the marks of each question will be allocated to questions on a short mini-passage taken from a relevant magazine or textbook article. You should expect such a passage to contain both familiar and new material and its questions to test both your knowledge of the topic and your comprehension of the mini-passage itself. The following section contains a checklist and a practice question for each of the four topics, look only at the part that covers your chosen topic and ignore the other three!

✳ Checklist for topic A – Astrophysics

Before attempting the following practice question on the astrophysics topic, check that you:

- ❏ have compared the use of photographic emulsions and charge-coupled devices in recording star images
- ❏ appreciate the effect of grain and pixel size on both their sensitivity and sharpness
- ❏ understand the importance of efficiency and linearity of response
- ❏ know the effects that the Earth's atmosphere has on the different radiations passing through it

❑ appreciate the benefits of observing stars using orbiting telescopes such as IRAS, COBE and Hubble

❑ know that the distance to even the closest star is very large

❑ are familiar with the use of the light year as a unit of length

❑ can explain fully the use of annual parallax to measure the distances to nearby stars

❑ can sketch the energy distribution graphs for stars having different temperatures and know the effect these have on colour

❑ appreciate that λ_{max} is not the maximum wavelength but the wavelength at which the emitted radiation has the greatest intensity

❑ can use Wien's law to find the surface temperature of a star once it's value of λ_{max} is known

❑ appreciate that surface temperatures of stars range from near absolute zero (λ_{max} in radiowaves) to 10^7 K (λ_{max} in X-rays)

❑ understand the difference between intensity and luminosity

❑ know how luminosity is calculated once distance and intensity have been measured

❑ appreciate how luminosity depends on surface temperature and surface area

❑ can sketch the Hertzsprung–Russell (H–R) diagram, showing the positions of the main sequence, red giants and white dwarfs, appreciating that temperature decreases towards the right

❑ realise that the H–R diagram is plotted using data obtained from nearby stars

❑ know how the H–R diagram is used to estimate the luminosity of a more distant main sequence star once its surface temperature has been calculated from its λ_{max}

❑ appreciate how this estimated luminosity and the measured intensity are used to determine how far away the more distant stars are and why annual parallax cannot be used to do this

❑ know that a Cepheid variable star is one whose brightness varies with a period that depends on its luminosity

❑ understand the use made of Cepheid variable stars in finding the distances to nearby galaxies

❑ appreciate that stars began as clouds of hydrogen gas that were pulled together by gravitational forces

❑ know that gravitational collapse results in increased temperature and can lead to 'hydrogen burning' where hydrogen fuses together to form helium

❑ can calculate the amount of energy released during the fusion process using $\Delta E = c^2 \Delta m$

❑ appreciate that the fusion process sets up an outward radiation pressure whose forces, in a stable star, balance the inward gravitational forces

❑ know that main sequence stars are in their 'hydrogen burning' stage

❑ understand that the more massive main sequence stars will have greater gravitational forces so will reach higher temperatures and be on the left of the H–R diagram

❑ appreciate that white dwarfs are hot (white) but have a low volume (dwarf) so that their surface area and, hence, luminosity are also low

❑ know that all white dwarfs are less than about 1.4 solar masses

□ appreciate that red giants are cool (red) with a large volume (giant) so that their surface area and, hence, luminosity are also high

□ know that all red giants are between 0.4 and 8 solar masses

□ understand that a supernova results from the shock wave, created by the rapid implosion of giant stars of more than 8 solar masses, which blows away the original star's outer layers into interstellar space

□ appreciate that a supernova explosion leaves behind a core remnant, and that remnants greater than about 1.4 solar masses form neutron stars whereas those greater than about 2.5 solar masses form black holes

□ know that neutron stars have extremely high densities

□ understand that a neutron star emits a beam of radio waves that sweeps across space as the star rotates and how this leads to its detection as a pulsar

□ appreciate that a black hole is so dense that no radiation can escape from it

Practice question for topic A – Astrophysics

Answers to this question, together with explanations, are in the Answers section which follows Chapter 6.

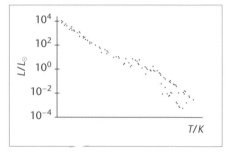

Fig 3.1

The following is a typical assessment question on the astrophysics topic. Attempt this question under similar conditions to those in which you will sit your actual test.

(a) Define power. [1]
State an appropriate unit for power. [1]
Express this unit in terms of base units. [2]

(b) Figure 3.1 shows a Hertzsprung–Russell diagram showing the main sequence. Luminosity of the Sun = $L_\odot$.
Draw a circle on the diagram showing the region where the Sun is located. Label this circle S.
Draw another circle showing the region where the most massive main sequence stars are located. Label this circle M. [2]
Indicate on the temperature axis the approximate temperatures of the coolest and of the hottest stars. [2]
Explain why large mass stars spend less time than the Sun on the main sequence. [2]
The luminosity of the Sun is 3.9×10^{26} W. Calculate the rate at which mass is being converted to energy in the Sun. [3]

(c) Charge coupled devices can have an efficiency as great as 70% compared with photographic film which has an efficiency of less than 5%. State two advantages of this greater efficiency. [2]
Explain why astronomical telescopes are sometimes launched into space. [2]

(d) Observations with a radio telescope in 1967 detected signals from a mysterious source which was called a pulsar. What type of star is a pulsar? [1]
What was unusual about the signals? [2]
Pulsars emit radio waves continuously. Explain why the signals detected on Earth are not continuous. You may be awarded a mark for the clarity of your answer. [3]

(e) Read the short passage below and answer the questions about it.

Cepheid variables are faint red giants whose brightness changes periodically. Their periodic changes in luminosity are the result of periodic pulsations of their giant bodies. A simple relationship exists

between the periods of these pulsations and the luminosities of the stars. The greater the luminosity, the longer the period of pulsation. This relationship has proved very useful for measuring the distances of stars which are too far away to show a parallax displacement. By measuring the pulsation period of a star its luminosity can be determined. This, combined with a measurement of the intensity at the Earth's surface, enables the distance to the star to be calculated.
[Adapted from *The Creation of the Universe* by George Gamow]

What is meant by the following terms used in the passage?

    Red giants
    Parallax displacement   **[3]**

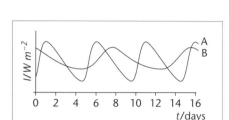

Fig 3.2

The curves in Figure 3.2 are plots of intensity against time for two Cepheid variable stars, A and B. These are known as light curves.
Estimate the period of the pulsations of star A.   **[1]**
What can be deduced about the luminosity of star B?   **[1]**
Since the average intensities of stars A and B are similar, what can be deduced about the distances of the two stars from the Earth? Give the reasoning which led to your answer.   **[2]**
Name the two forces, one which causes a star to contract and one which causes it to expand, which must be repeatedly out of balance in a Cepheid variable.   **[2]**

(Total 32 marks)
(*Edexcel Unit Test PHY3, June 2001, Q. 1*)

 Checklist for topic B – Solid materials

Before attempting the following practice question on the solid materials topic, check that you:

❏ can sketch force–extension graphs for copper, mild and high carbon steel and rubber

❏ understand the meanings of elastic limit, limit of proportionality and yield point

❏ appreciate the differences between elastic and plastic behaviour

❏ have learnt a statement of Hooke's law

❏ know that when a material is stretched elastically, elastic strain energy is stored

❏ appreciate that work done is the area under a force–extension graph

❏ know that a lot more energy is used to stretch a copper wire plastically than elastically but that only the elastic portion of this is recoverable

❏ appreciate the difference between tough and brittle materials

❏ know that a ductile material can be pulled out into wires while a malleable material can be beaten into shape

❏ can calculate values of stress, strain and the Young modulus

❏ know that a material with a large Young modulus is stiff whereas one with a small Young modulus is flexible

❏ can sketch stress–strain graphs for copper, mild and high carbon steel and rubber

❏ appreciate that the Young modulus is the slope of the initial linear section of a stress–strain graph

❏ understand the meanings of yield stress and ultimate tensile stress and know that the latter determines if a material is strong or weak

- ❏ appreciate that the area under a stress–strain graph gives the energy stored per volume
- ❏ can explain the temperature increase experienced by a repeatedly stretched and relaxed rubber band by referring to its hysteresis behaviour
- ❏ appreciate that the area between the loading and unloading curves is the energy absorbed by the rubber during each stretching and relaxing cycle
- ❏ know that most metals are polycrystalline
- ❏ understand what is happening to the atoms of a metal during both elastic and plastic stretching
- ❏ can describe how the presence of edge dislocations and slip planes reduce the strength of a crystal
- ❏ appreciate that work hardening produces lots of dislocations that by getting tangled together makes it harder for plastic deformation to take place and strengthens the material
- ❏ know that dislocations cannot move beyond grain boundaries so metals with small crystals are stronger than metals with larger crystals
- ❏ understand the differences between the processes of annealing, quench hardening and tempering
- ❏ appreciate that the strain of some materials gradually increases over time (creeps) even though the stress remains constant
- ❏ know that cyclic loading and unloading can produce cracking and lead to fatigue failure
- ❏ appreciate that stress is greatest at the tip of a crack where the area is least
- ❏ know that polymers consist of long chain molecules that are either randomly arranged (amorphous) or partly random and partly ordered (semi-crystalline)
- ❏ understand that thermoplastics have weak bonds between adjacent long chains so they soften and can be injection moulded at high temperatures
- ❏ appreciate that thermosets are rigid whatever the temperature due to strong permanent crosslinks
- ❏ know some of the uses made of thermoplastics and thermosets
- ❏ can explain the shape of the stress–strain graph for rubber with reference to the uncoiling and stretching of its long chain molecules
- ❏ understand that a composite material combines two or more materials to make the best use of their individual properties and know a number of examples
- ❏ appreciate that concrete is strong in compression but weak in tension due to cracking
- ❏ know how pre-stressed reinforced concrete uses tensioned steel cables to keep the concrete in compression
- ❏ can apply the principle of moments studied in Unit 1 to systems involving non-parallel forces

Practice question for topic B – Solid materials

Answers to this question, together with explanations, are in the Answers section which follows Chapter 6.

The following is a typical assessment question on the solid materials topic. Attempt this question under similar conditions to those in which you will sit your actual test.

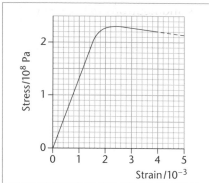

Fig 3.3

(a) Define work. **[1]**

State an appropriate unit for work. **[1]**

Express this unit in terms of base units. **[2]**

(b) State Hooke's law. **[2]**

The graph in Figure 3.3 shows the stress–strain relationship for a copper wire under tension.

Use the graph to determine:

 the ultimate tensile stress for copper

 the Young modulus of copper **[3]**

A copper wire of cross-sectional area 1.7×10^{-6} m^2 and length 3.0 m is stretched by a force of 250 N.

Will the behaviour of the wire at this point be elastic or plastic? Justify your answer. **[2]**

Show this point on the stress-strain graph in Figure 3.3. Label it P. **[1]**

Calculate the extension of the wire. **[2]**

(c) Explain with the aid of a diagram what is meant by an edge dislocation. **[2]**

Describe how the presence of dislocations can reduce the risk of metals failing by cracking. You may be awarded a mark for the clarity of your answer. **[3]**

(d) Sketch a force–extension graph for natural rubber showing its behaviour for both increasing and decreasing force. **[2]**

Use your graph to explain why a rubber band, which is repeatedly stretched and relaxed, becomes noticeably warmer. **[2]**

(e) Read the short passage below and answer the questions about it.

The outer layer of a human tooth is made of enamel and is the hardest tissue in the body. It is a typical ceramic with a high compressive strength, low tensile strength and a high Young modulus. It is brittle and consists of long crystals of calcium phosphate set vertically on the surface of the underlying dentine. Dentine is the main structural material of a tooth. It is a composite material consisting of needle shaped crystals in a collagen fibre matrix. Dentine has a much lower Young modulus than enamel and is tough. False teeth (dentures) are made from PMMA (polymethyl methacrylate), an amorphous polymer with a glass transition temperature of 110 °C.

What is meant by the following term used in the passage?

 Composite material **[2]**

Draw labelled diagrams to show the difference in molecular structure of a crystalline material and an amorphous polymer. **[3]**

The graph in Figure 3.4 shows the behaviour of enamel under tension. Add to the graph:

(i) a line labelled E to show the behaviour of enamel under compression,

(ii) a line labelled D to show the behaviour of dentine under tension.

[4]

(Total 32 marks)

(Edexcel Unit Test PHY3, June 2001, Q. 2)

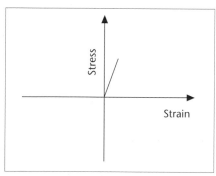

Fig 3.4

Checklist for topic C – Nuclear and particle physics

Before attempting the following practice question on the nuclear and particle physics topic, check that you:

- know that the nucleus is made of protons and neutrons that are very close together
- appreciate that the protons are positively charged and repel each other strongly
- understand that there must be another force holding them together, the strong nuclear force
- know that the strong nuclear force attracts both protons and neutrons together but has a much shorter range than the electrostatic repulsion between protons
- appreciate that the radius r of a nucleus increases with nucleon number A, such that $r = r_0 A^{1/3}$ where r_0 is the radius of a proton
- understand why the density of nuclear matter is very much larger than that of the material containing it
- can sketch a graph showing how the number of neutrons N varies with the number of protons Z for all nuclides
- know that radioactive decay tends to bring a nuclide closer to the N–Z trend line
- appreciate the effects of α, β^- and β^+ decays on a nuclide's position
- know what is meant by a decay chain and can plot an N–Z curve for any given decay chain
- understand the principles of radioactive dating using both carbon and uranium
- can sketch the energy spectrum for a typical α particle decay
- understand why all α particles from the same decay have the same energy
- can sketch the energy spectrum for a typical β^- decay
- appreciate that in β^- decay a neutron in the nucleus splits into a proton and an electron
- understand how the emission of β^- particles with a range of energies from the same decay suggests that a further particle, an antineutrino, is also emitted
- appreciate why the antineutrino must be neutral and of very low mass
- know that in β^+ decay a neutrino is also emitted
- know the principle of conservation of mass-energy
- can calculate the decrease in nuclear mass (in u) for a nuclear decay and express this as an energy in MeV (1 u = 930 MeV)
- understand what is meant by binding energy per nucleon and how it relates to nuclear stability
- are aware of the existence of antimatter and can name a number of particle-antiparticle pairs
- appreciate that when a particle meets its antiparticle they annihilate each other such that both particles disappear and energy is released as electromagnetic photons
- can calculate the energy released from the rest masses and energies of the particles involved
- understand that a fundamental particle is one that cannot be split into anything smaller
- appreciate that particles can be classified as either leptons (light) or hadrons (heavy) and can name several examples of each type
- know that leptons are fundamental particles while hadrons consist of smaller fundamental particles called quarks

Unit 3

- ❑ appreciate that there are six different types of quark, each with its associated antiquark
- ❑ know that individual quark properties are referred to as flavour
- ❑ know that there are three types of hadrons: mesons consisting of a quark-antiquark structure, baryons consisting of three quarks and antibaryons consisting of three antiquarks
- ❑ have learnt the quark structure of a neutron and a proton
- ❑ understand that charge, lepton number and baryon number are conserved in all particle interactions
- ❑ can apply these conservation laws to any given particle interaction
- ❑ know the four ways in which fundamental particles may interact
- ❑ appreciate that the gravitational interaction occurs with all particles having mass
- ❑ know that the electromagnetic interaction occurs with all charged particles
- ❑ appreciate that the weak interaction occurs with all particles while the strong interaction only occurs between quarks
- ❑ know that the weak interaction is the only one that can produce a change in quark flavour
- ❑ understand that during interactions, set amounts (quanta) of energy are exchanged
- ❑ can name the exchange particles associated with each of the four fundamental interactions
- ❑ can draw and interpret simple Feynman diagrams

Practice question for topic C – Nuclear and particle physics

Answers to this question, together with explanations, are in the Answers section which follows Chapter 6.

The following is a typical assessment question on the nuclear and particle physics topic. Attempt this question under similar conditions to those in which you will sit your actual test.

(a) Define density. **[1]**

Radius of a gold atom $\approx 10^{10}$ m   Radius of a gold nucleus $\approx 10^{-15}$ m

Show that   (volume of a gold atom)/(volume of a gold nucleus) $= 10^{15}$

The density of gold is 19 000 kg m^{-3}. Estimate the density of a gold nucleus.

What assumption have you made in obtaining your answer? **[3]**

(b) Sketch graphs showing the energy spectra (i.e. number of particles against their kinetic energy) for

(i) a typical α particle decay,

(ii) a typical β^- decay. **[3]**

Complete the equation below showing the decay of a neutron to a proton. **[2]**

$n \rightarrow \quad p^+$

State the quark compositions of the neutron and the proton. Use these to explain why only the weak interaction can be responsible for this decay. You may be awarded a mark for the clarity of your answer. **[3]**

(c) The grid in Figure 3.5 shows the relationship between number of neutrons N and number of protons Z for some of the stable nuclides in the region $Z = 31$ to $Z = 45$.

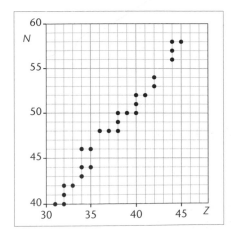

Fig 3.5

Strontium-90, $^{90}_{38}$Sr, is an unstable nuclide. It decays by β^- emission to an unstable isotope of yttrium. On the grid mark the position of $^{90}_{38}$Sr and this isotope of yttrium. **[2]**

$^{82}_{37}$Rb is another unstable nuclide. Mark the position of $^{82}_{37}$Rb on the grid. By what means would you expect $^{82}_{37}$Rb to decay? **[2]**

(d) The following strong interaction has been observed.

$$K^- + p \rightarrow n + X$$

The K⁻ is a strange meson of quark composition ūs.
The u quark has a charge of +2/3.
The d quark has a charge of −1/3.
Deduce the charge of the strange quark. **[1]**
Use the appropriate conservation law to decide whether particle X is positive, negative or neutral. **[2]**
Is particle X a baryon or a meson? Show how you obtained your answer. **[2]**
State the quark composition of X. Justify your answer. **[3]**

(e) Read the short passage below and answer the questions about it.

In 1974 electron-positron collisions led to the discovery of the psi particle (Ψ). The Ψ is a meson of composition cc̄, that is it contains a charmed quark and a charmed antiquark. The c and c̄ move around one another, rather like the electron and proton in a hydrogen atom. A variety of orbitals of different energy are possible. If the c and c̄ orbit with high energy, they form a relatively heavy particle. A heavier version of the Ψ meson was soon discovered. It is written Ψ'. The Ψ' rapidly loses energy and decays to a Ψ and two pi mesons, π⁺ and π⁻. This is followed by the decay of the Ψ as its c and c̄ annihilate, their energy appearing as a positron and an electron.
[Adapted from *The Particle Explosion* by Close, Marten and Sutton]

List all the antiparticles listed in the passage. **[2]**
Although the Ψ contains charmed quarks, it has zero charm itself. Explain how this can be so. **[1]**
Describe how a charmed quark and a charmed antiquark can create a heavier version of the Ψ. **[1]**
The Feynman diagram in Figure 3.6 shows the decay of the Ψ'. Complete the diagram by identifying the particles A, B and C. **[2]**
What fundamental interaction is responsible for this decay? **[1]**
Identify the exchange particle involved. **[1]**

(Total 32 marks)
(Edexcel Unit Test PH3, June 2001, Q. 3)

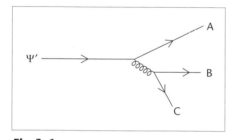

Fig 3.6

Unit 3

 Checklist for topic D – Medical physics

Before attempting the following practice question on the medical physics topic, check that you:

☐ understand the difference between diagnosis and therapy

☐ appreciate that radiation can mutate or kill living cells

☐ know that alpha radiation is stopped by the skin and beta radiation by a few centimetres of flesh

☐ appreciate why radioisotopes that emit only gamma radiation are generally preferred

☐ know the functions of the main parts of a gamma camera

☐ understand how, by injecting a known volume of radioactive tracer with a known activity into the blood stream, the volume of blood in a patient can be determined

❑ know that most radioisotopes used in medicine are prepared by neutron bombardment

❑ appreciate how different radioisotopes are selected for different tasks

❑ know that high energy (1.3 MeV) gamma rays from ^{60}Co are used to kill cancerous cells

❑ understand why ^{123}I (γ only) is used to monitor thyroid function whereas ^{131}I (β^- and γ) is used to destroy part of an over-active thyroid

❑ know that metastable radionuclides such as ^{99m}Tc are in a higher energy state than normal and decay to a more stable state by emitting gamma radiation

❑ understand the principle of an elution cell and how it provides a convenient source of gamma radiation

❑ appreciate that the amount of radioisotope remaining in the body is determined by a combination of how quickly it decays and how quickly the body excretes it

❑ know the meanings of radioactive and biological half-lives

❑ can use their values to calculate the effective half-life

❑ are aware of the basic principles of radiological protection

❑ know that X-rays are produced when high speed electrons are suddenly stopped

❑ can draw and label the structure of a typical X-ray tube and know why its anode is rotated

❑ appreciate that the efficiency of an X-ray tube is very low

❑ understand how the maximum energy (in eV) of emitted X-rays depends on the accelerating voltage

❑ know that X-rays with energies around 100 keV are used for diagnosis

❑ appreciate that bone absorbs a lot more of these X-rays than tissue or air due to its larger proton number

❑ know that a photographic plate, on the other side of the patient, detects non-absorbed X-rays

❑ appreciate how using a point source of X-rays and a lead anti-scatter grid improves image sharpness

❑ understand why the intensity of X-rays from a point source follows an inverse square law

❑ know that X-rays with energies around 1 MeV are used for therapy

❑ appreciate that the absorption of these X-rays is less dependent on proton number and so can be used to treat tissue as well as bone

❑ understand the need for alignment devices and why a multiple-beam rotational treatment method is used

❑ appreciate the problems associated with using X-ray doses that are either too low or too high

❑ know that ultrasound is very high frequency longitudinal waves

❑ appreciate that ultrasonic diagnosis relies on waves being reflected from boundaries along its path – the sonar principle

❑ know that the amount of reflection depends on the difference in the acoustic impedance of each material

❑ can calculate the reflection coefficient for a given boundary

❑ have learnt the basic principles of the A- and B-scan methods for ultrasonic diagnosis

❑ appreciate that the probe used functions as both transmitter and receiver

- know why a coupling medium (gel) is used between the probe and the skin
- understand how time delays are used to calculate depths of boundaries
- have learnt and can use the equation $c = f\lambda$
- appreciate that small wavelengths suffer less diffraction and so give better resolution although they are also more readily absorbed and so the reflected signal is weaker
- have compared ultrasound and X-ray diagnostic techniques

Practice question for topic D – Medical physics

The following is a typical assessment question on the medical physics topic. Attempt this question under similar conditions to those in which you will sit your actual test.

> Answers to this question, together with explanations, are in the Answers section which follows Chapter 6.

(a) Define electrical potential difference. **[1]**
State an appropriate unit for potential difference. **[1]**
Express this unit in terms of base units. **[2]**

(b) Explain why the effective half-life of a radioisotope administered to a patient is less than the half-life due to radioactive decay. **[1]**
^{131}I has a radioactive half-life of 8 days and a biological half-life of 21 days. Calculate the effective half-life of ^{131}I. **[2]**
After how many days will the fraction of a sample of ^{131}I remaining in the body be 1/8 of the administered dose? **[2]**
Give one reason why gamma-emitting radionuclides are preferred for tracer studies. **[1]**
State another property (other than half-life) that is important when selecting an appropriate gamma-emitting radionuclide for diagnostic purposes. **[1]**

(c) Explain why a coupling medium between the transducer and the body surface is necessary when carrying out an ultrasound scan. You may be awarded a mark for the clarity of your answer. **[3]**
Suggest an appropriate substance for use as a coupling medium. **[1]**
Figure 3.7 shows an A-scan trace on an oscilloscope. The pulses represent reflections from opposite sides of the head of a fetus.
The time base of the oscilloscope is set at 50 μs div^{-1}. The speed of sound in the fetal head is 1.5×10^3 m s^{-1}. Calculate the size of the head of the fetus. **[4]**

(d) Figure 3.8 shows part of a diagnostic X-ray tube.
Suggest an appropriate operating voltage for this tube. **[1]**
Why is the anode rotated? **[1]**
Why is the X-ray tube evacuated? **[1]**
Suggest an appropriate material for the outer case. **[1]**

(e) Read the short passage below and answer the questions about it.

Attenuation is the reduction in intensity of a beam as it travels. X-ray beams are usually heterogeneous, that is they contain X-rays of many different wavelengths. In passing through a medium the different wavelengths are attenuated by different amounts. Longer wavelength (lower energy) X-rays are attenuated more than shorter wavelength ones. After passing through a filter the remaining X-rays therefore have a higher average energy and are relatively more penetrating. A more penetrating beam is said to be of better quality. As a heterogeneous

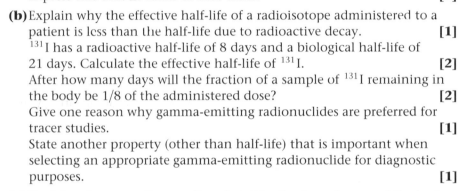

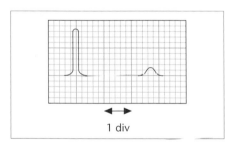

Fig 3.7

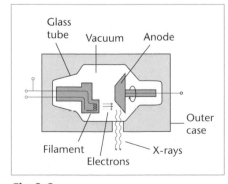

Fig 3.8

Unit 3

beam passes through a medium its quality gradually increases. This process is described as hardening. The quality of an X-ray beam may be improved by either increasing the tube voltage or using a filter.
[Adapted from *Medical Physics Imaging* by J~ Pope]

State the meaning of the following terms used in the passage.

    Heterogeneous
    Hardening   **[2]**

Figure 3.9 shows the distribution of different wavelength X-rays in a typical X-ray beam.
Add to the graph to show the possible distribution of X-rays after passing this beam through a filter.   **[3]**
Why is the X-ray beam relatively more penetrating after it has been filtered?   **[2]**
Suggest why it is beneficial to the patient to filter the beam.   **[2]**

(Total 32 marks)
(Edexcel Unit Test PHY3, June 2001, Q. 4)

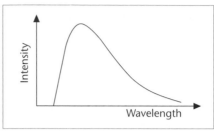

Fig 3.9

The AS practical test

Introduction

This test examines your practical laboratory skills such as how you plan and perform experiments and how you analyse results and draw conclusions. It is based on the content of Units 1 and 2. The test consists of two questions each one lasting 40 minutes, although the apparatus may only be used for the first 35 of these. There is a further 10 minutes writing up time at the end giving a total test length of 90 minutes. Each question is worth 24 marks. Question 1 consists of a number of short practical exercises mainly involving setting up and using apparatus and recording observations. Question 2 concentrates on planning and evaluation. Neither of the questions requires the use of datalogging apparatus although you may be asked to explain how to set up and use such a device. The following section gives advice on how to maximise your AS practical test mark, together with a sample question of each type. Where possible, you should also practise these questions using the apparatus, which is listed in the questions.

Advice on tackling the AS practical test

General

- make sure that you know how to use all standard apparatus met with in Units 1 and 2 such as vernier callipers, micrometers, analogue and digital electrical meters
- if you work on class practicals in pairs, take it in turns to assemble and use the apparatus
- treat each of your class practicals as a test to develop good habits throughout the course
- ask your teacher to arrange for you to attempt a number of practice questions under test conditions
- remember that the examiner is not in the room watching you and no video record is kept of what you do, so if what you are doing is important then write it down
- although the examiner is not there, the supervisor has to record certain measurements on the front of your test paper so that the examiner can compare your values with these and award marks accordingly – unfortunately these are added after you have finished the test!
- pay attention to significant figures – practical tests are the only ones where too many or too few are penalised – as a general rule, it is best to keep to two or three

Methods

- the question tells you what to do so don't write out a general method
- concentrate on what you must do to achieve accurate results but still only describe this if asked to do so

- remember to use diagrams to help with any descriptions
- diagrams should be drawn carefully using a straight edge and labelled
- when labelling distances, make sure the labels accurately show the correct end-points
- if apparatus has to be vertical then align it with a door or window frame
- if apparatus has to be horizontal then check that each end is the same height above the bench

Measurements

- make sure you follow all instructions carefully
- always check such instruments as vernier callipers and micrometers for zero errors and tell the examiner that you have done so
- take measurements to the smallest division of the instrument used; e.g. 0.1 mm with vernier callipers
- always give the correct units for all your measurements
- avoid parallax errors by having the eye positioned adjacent to the reading
- show all the measurements that you have to take; avoid doing sums in your head and just writing down the answer; for example, when finding the mass of water in a container
- repeat readings should be taken and must be written down even if they are all identical
- tabulate any series of corresponding readings and include the units in the table headings
- leave apparatus set up so that further results can be taken if shown to be needed; for example, by a graph

Graphs

- always use sensible scales that are easy to use and easy to follow – don't use steps of 3, 6, 7, 9 etc
- scales must allow all points to be plotted and the plotted points must occupy at least half of the grid
- label both axes with both quantities and units
- plot points accurately in pencil using either '×' or '⊙' – your plotting will be checked especially those points that are furthest from your line
- fill in any large gaps by taking further measurements
- practise drawing best-fit straight lines and curves well before you sit the exam
- a long (30 cm) clear plastic ruler is essential for judging best-fit straight lines – make sure you have one
- recheck points that are furthest from your line and adjust if found to be incorrect

Calculations

- these may involve the gradient *m* or the intercept *c* of any straight line graph that you have plotted
- use as large a triangle as possible when calculating a gradient

- the gradient at a point on a curve is found by drawing a tangent to the curve at that point
- remember that most gradients have units, those of y divided by those of x
- a line sloping down from left to right has a negative gradient
- the units of the intercept is the same as those of y
- give all calculated answers to the same number of significant figures as your measurements

Uncertainties

- only work out uncertainties if and where the question tells you to do so, therefore for most of your measurements you won't have to bother with uncertainties
- for a single measurement take the smallest division of the instrument as its uncertainty e.g. a length might be 19 ± 1 mm using a metre rule and 18.8 ± 0.1 mm using vernier callipers
- practise finding percentage uncertainties (equation given in test); e.g. 5.3 and 0.53% for above lengths
- for a set of repeated readings take half the spread of the readings as the uncertainty of the average value e.g. 1.25, 1.29, 1.28 and 1.26 mm giving a spread of 0.04 mm so average value = 1.27 ± 0.02 mm with a percentage uncertainty of 1.6%
- for measurements, such as starting and stopping a timer, where your own error adds significantly to the uncertainty, always take several measurements and use their range to get the uncertainty
- add the uncertainties of any measurements that are either added or subtracted
- add the percentage uncertainties of any measurements that are multiplied or divided
- multiply the percentage uncertainty by any power to which the measurement is raised e.g. percentage error in r^3 is $3 \times$ percentage uncertainty in r
- use $100 \times$ difference/average value to calculate the percentage difference between two values

 Sample AS practical questions

Apparatus needed: A4 piece of card about 1 mm thick, half-metre rule, micrometer, electronic top pan balance

1 (a) (i) Taking care not to damage the card supplied, determine average values for the length l, the width w and the thickness t. Explain why it is necessary to take a number of values in order to determine accurate values for the above quantities. **[6]**

> $l = 297$ mm, 297 mm   average $l = 297$ mm
> $w = 210$ mm, 210 mm   average $w = 210$ mm
> > l and w within ± 2 *mm* of supervisor and to mm (or better) precision ✓
> > both repeated and averaged ✓
> no zero error on micrometer
> $t = 0.97$ mm, 0.94 mm, 0.98 mm, 0.95 mm   average $t = 0.96$ mm
> > t within ± 0.03 mm of supervisor and to 0.01 mm (or better) precision ✓

averaged from at least two readings or ±0.05 mm from at least two readings ✓
reference to zero error or at least four readings of t ✓
e.g. to eliminate anomalous/rogue measurements or card thickness may vary
suitable explanation ✓

(ii) Using the top pan balance, measure the mass of the card and hence find a value for the density of the material of the card. The value you have obtained for the average thickness of the card is not necessarily the best average value. Explain how you could obtain a better average value for the thickness. You may assume that additional apparatus is available.

[4]

mass of card = 38.7 g
density = mass/volume = 38.7×10^{-3} kg/(0.297 m × 0.210 m × 0.96×10^{-3} m) = 646 kg m^{-3}
correct density calculation and unit ✓
2 or 3 s.f. from correct calculation ✓
e.g. measure total thickness of 20 pieces and average
use longer reach micrometer/cut card to get at centre
two appropriate explanations ✓✓

(b)(i) Set up the circuit shown in Figure 3.10. Before you close the switch, have your circuit checked by the supervisor. You will be allowed a short time to correct any faults, but if you are unable to set up the circuit the supervisor will set it up for you. You will only lose 2 marks for this. **[2]**

circuit set up correctly without help [and e.m.f. in (ii) not 0 V!] ✓✓

Apparatus needed: 1.5 V dry cell, switch, digital voltmeter, 4.7 Ω resistor in holder, five connecting leads

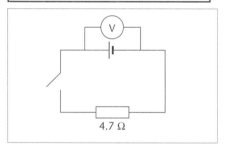

Fig 3.10

(ii) You may assume that the voltmeter is an ideal voltmeter which takes no current. Use your circuit to determine the e.m.f. ε of the cell and the potential difference V across the 4.7 Ω resistor. Leave the switch open after you have completed your readings. **[2]**

ε = 1.48 V and V = 1.26 V
sensible value of ε to 0.01 V (or better) precision ✓
sensible value of V to 0.01 V (or better) precision ✓

(iii) Calculate the current I through the resistor. Hence calculate the internal resistance r of the cell. **[3]**

$I = V/R$ = 1.26 V/(4.7 Ω) = 0.268 A
correct calculation of I with unit and to at least 2 s.f. ✓
$r = (\varepsilon - V)/I$ = (1.48 V − 1.26 V)/(0.268 A) = 0.82 Ω
substitution into correct formula ✓
correct calculation with unit and to 2 or 3 s.f. ✓

Apparatus needed: 200 cm^3 of water at room temperature in 250 cm^3 beaker, 100 cm^3 measuring cylinder, expanded polystyrene cup, −10 °C to +110 °C thermometer, plastic stirrer, card telling you the mass of the 10 washers, paper towels for mopping up, access to 10 mild steel washers (diameter 25 mm, thickness 1 mm) of known mass, tied together to a length of thin string and pre-heated in boiling water

(c) (i) Place 50 cm^3 of water at room temperature in the polystyrene cup. Record the temperature θ_1 of the water. The supervisor has placed 10 washers tied together with string in a beaker of boiling water. Using the string, remove the washers from the beaker and transfer them to the polystyrene cup. Record the highest steady temperature θ_2 reached by the water. Calculate the specific heat capacity c_s of mild steel given that

$$c_s = m_w c_w (\theta_2 - \theta_1)/[m_s(\theta_3 - \theta_2)]$$

where
m_w = mass of water = 0.050 kg
c_w = specific heat capacity of water = 4200 J kg^{-1} K^{-1}
m_s = mass of 10 washers = 0.039 kg
θ_3 = initial temperature of washers = 100 °C **[5]**

$\theta_1 = 17.4\ ^\circ C$ and $\theta_2 = 23.1\ ^\circ C$

    sensible θ_1 and θ_2 recorded with unit ✓

$c_s = 0.050\ \text{kg} \times 4200\ \text{J kg}^{-1}\ \text{K}^{-1} \times (23.1 - 17.4)\ ^\circ C/[0.039\ \text{kg} \times$
$100 - 23.1)\ ^\circ C] = 399\ \text{J kg}^{-1}\ \text{K}^{-1}$

    correct substitution (consistent units) ✓

    correct calculation with unit ✓

    good value i.e. 400 ± 100 ✓✓ (or 400 ± 200 ✓)

(ii) State two sources of error in this experiment. **[2]**

e.g. loss of heat from washers during transfer

some water transferred with washers

some heat gained by thermometer/cup

heat lost from cup to surroundings

    two appropriate sources of error ✓✓

(Edexcel Unit Test PHY3, May 2001, Q. 2A)

2 (a) The apparatus shown in Figure 3.11 has been set up for you. Add masses to the mass hanger until it is clear that the trolley accelerates across the table. Record the total mass m used to accelerate the trolley.
Determine the average time t for the trolley to travel a distance $x = 0.500$ m from rest when accelerated by this mass.
Calculate the acceleration of the trolley given that $a = 2x/t^2$ **[4]**

$m = 40$ g (i.e. hangar + 30 g)

$t = 3.7$ s, 3.8 s, 3.7 s, 3.6 s   average $t = 3.7$ s

    t to at least 0.1 s precision, repeated and averaged, with unit ✓

    average from at least 3 readings ✓

    value $= 3.5 \pm 1.5$ s ✓

$a = 2x/t^2 = 2 \times 0.500\ \text{m}/(3.7\ \text{s})^2 = 0.073\ \text{m s}^{-2}$

    correct calculation with unit to at least 2 s.f. ✓

(b) Explain with the aid of a diagram how you ensured that the trolley travelled a distance of 0.500 m in the measured time. **[3]**

see Figure 3.12 (good labelled diagram sufficient for all 3 marks)

    diagram clearly showing distance travelled ✓

    using same point on trolley ✓

    good method e.g. eye position shown level with each position ✓

(c) Applying Newton's second law to this system gives

$(M + m)a = mg - F$

where M = the mass of the trolley and its load = 2.42 kg

    F = the frictional force opposing the motion of the system

    g = the gravitational field strength

Use your results from part (a) to calculate a value for F. **[3]**

$F = mg - (M + m)a = (0.04\ \text{kg} \times 9.81\ \text{N kg}^{-1}) - [(2.42 + 0.04)\ \text{kg} \times 0.073\ \text{m s}^{-2}] = 0.21\ \text{N}$

    correct rearrangement of equation ✓

    correct substitution using SI units ✓

    correct calculation with unit to at least 2 s.f. ✓

Apparatus needed: loaded trolley of known mass (2 to 3 kg) joined with about 1 m of thin string to a 10 g mass hanger suspended over a pulley as shown in Figure 3.11; nine 10 g slotted masses, metre rule, digital stopwatch

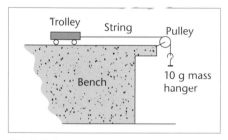

Fig 3.11

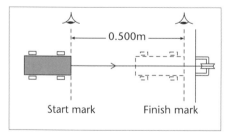

Fig 3.12

Unit 3

(d) Repeat the experiment with a larger value of *m* in order to calculate a second value for *F*. **[4]**

> Using *m* = 90 g
> *t* = 2.0 s, 1.9 s, 2.0 s, 2.1 s   average *t* = 2.0 s
> $a = 2x/t^2 = 2 \times 0.500$ m$/(2.0$ s$)^2 = 0.25$ m s^{-2}
> $F = mg - (M + m)a = (0.09$ kg $\times 9.81$ N kg$^{-1}) - [(2.42 + 0.09)$ kg $\times 0.25$ m s$^{-2}] = 0.26$ N
>> chosen *m* at least 30 g larger than previous value   ✓
>> *t* to at least 0.1 s precision, repeated and averaged, with unit   ✓
>> average from at least 3 readings   ✓
>> correct calculation of *F* with units to at least 2 s.f.   ✓

(e) Calculate the percentage difference between your two values of *F*. Comment on the extent to which the value of *F* may be regarded as constant if it is assumed that experimental errors are in the region of 10%. **[2]**

> Percentage difference = $100 \times (0.26 - 0.21)$ N$/(0.235$ N$) = 21\%$
> as this is >10%, *F* cannot be regarded as constant
>> correct calculation of percentage difference   ✓
>> sensible conclusion   ✓

(f) The equation in part (c) may be investigated by plotting a graph of $(M + m)a$ against *m*.

  (i) Explain carefully how you would carry out the experiment to plot this graph.

  (ii) Sketch the graph you would expect to obtain if the force *F* were constant.

  (iii) State the values you would expect to obtain for both the gradient and the intercept on the vertical axis. **[8]**

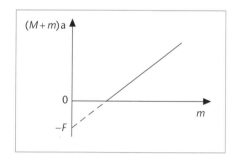

Fig 3.13

  (i) keep *M* constant   ✓
       use range of values of *m*   ✓
       measure corresponding values of *t*   ✓
       calculate corresponding values of *a*   ✓

  (ii) From part (c) $(M + m)a = gm - F$
        compare this with $y = mx + c$ to give graph shown in Figure 3.13

   Sketch graph as shown in Figure 3.13
       graph of $(M + m)a$ against *m* with axes labelled   ✓
       correct straight line   ✓
       with extrapolation to vertical axis   ✓

  (iii) gradient = *g*   ✓
        intercept with vertical axis = $-F$ ✓

  (Edexcel Unit Test PHY3, May 2001, Q. 2B)

4 Waves and our universe

Part ❶ Oscillations and waves

 Introduction

Waves are the result of a disturbance such as the oscillation of an object. The simplest form of a wave is one represented by a sine or cosine function, indicating that the oscillator producing it moves with simple harmonic motion. The to-and-fro motion of such an oscillator is similar to the side-to-side motion of a body moving around a circular path at a constant speed. As you study oscillations and waves, you learn why a body following a circular path continues to move at a constant speed despite the fact that it is accelerating. You find that a simple harmonic oscillator has a constant period and can be used for accurate timing. You investigate how the period of a pendulum depends on its length and use the resulting relationship to calculate a value for the acceleration of gravity. You study the effect of damping on the resonance curve of a forced oscillator and meet with everyday situations involving resonance. You become familiar with terms such as amplitude, speed, wavelength, frequency and phase, which are used to describe a wave. You discover the difference between progressive and stationary waves, between longitudinal and transverse waves and between mechanical and electromagnetic waves. You find that waves can reinforce or cancel depending on their phase difference and you measure wavelength from a number of superposition experiments.

Unit 4

 Things to understand

Circular motion

- angular speed ω describes the circular motion of a body in terms of the rate of change of the angle $\Delta\theta$ at the centre of its path, $\omega = \Delta\theta/\Delta t$

- this angle is measured in radians (rad) and angular speed in rad s^{-1}

- there are 2π radians in a whole circle, so dividing 2π by ω gives the period T, the time taken to complete one revolution

- the period can also be found by dividing the distance around the circular path (circumference = $2\pi r$) by the linear speed v of the object

- bodies can have the same angular speed but different linear speeds depending on how far they are from the centre of the circle (Figure 4.1)

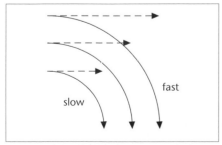

Fig 4.1 *The speed increases with path radius for objects with the same angular speed*

Unit 4

- a body moving around a circular path is continually changing its direction of motion
- the body's velocity (a vector) is continually changing although its speed (a scalar) remains constant
- since its velocity is continually changing, the body is always accelerating
- the acceleration is towards the centre of the circular path (centripetal acceleration)
- the acceleration of gravity is the centripetal acceleration for satellites orbiting the Earth
- a resultant force (centripetal force) must act towards the centre to produce this acceleration
- without this force, the body would continue in a straight line along a tangent
- the work done ($F\Delta x$) by a centripetal force is zero as there is no displacement ($\Delta x = 0$) in the direction of the force
- a centripetal force does not change the kinetic energy or speed of a body, only its direction of motion

Weightlessness

- you can only be weightless when you are a long way from all other masses
- if you were falling freely, you would experience apparent weightlessness as all parts of your body would then be accelerating downwards at the same rate
- astronauts in orbit around the Earth are continually accelerating towards its centre at a rate equal to the acceleration of gravity and experience apparent weightlessness

Oscillations

- many systems undergo to-and-fro movements called oscillations
- all oscillations eventually die away as energy dissipates to the surroundings
- the position of an oscillating body is given as its displacement (a vector) from its equilibrium position
- when at its equilibrium position, the resultant force on a body is zero
- displacements vary from positive to negative to positive during the course of one oscillation
- amplitude is the maximum displacement from the equilibrium position
- slow oscillations are usually described using their periods while for faster oscillations the frequency, the number of oscillations per second, is used

Simple harmonic motion

- Simple harmonic motion (s.h.m.) is an oscillation where the period remains constant even when the amplitude changes
- a pendulum and a mass-spring system both oscillate with s.h.m.
- the regularity of simple harmonic oscillators is used in many clocks
- plotting displacement against time for such an oscillator produces a sinusoidal graph

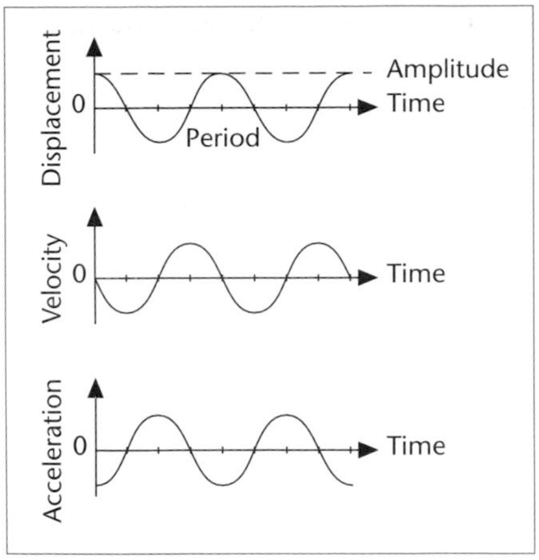

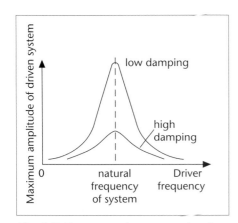

Fig 4.2 *Displacement, velocity and acceleration for s.h.m.*

- the velocity–time graph can be deduced from the variation of the gradient of the displacement–time graph
- the acceleration–time graph can be deduced from the variation of the gradient of the velocity–time graph
- velocity and acceleration also vary sinusoidally with time although all three quantities are out of step with each other (Figure 4.2)
- at the equilibrium position (zero displacement), velocity is a maximum (either positive or negative) and acceleration is zero (it has to be since resultant force is zero)
- positive and negative velocities show the body moving in opposite directions
- at either amplitude, velocity is zero (the body momentarily stops moving as it changes direction) and acceleration is a maximum (either positive or negative)
- positive and negative accelerations show the body accelerating in opposite directions
- acceleration is always directed towards the equilibrium position and increases with distance from it
- acceleration is proportional to displacement but in the opposite direction (acceleration and displacement always have opposite signs)
- the relationship $a \propto -x$ results in s.h.m. and is used to define it

Forced oscillations

- every system has its own natural frequency at which it oscillates
- a system oscillates with its largest amplitude (resonates) when forced to oscillate at its natural frequency
- resonance is sharper when damping is low (Figure 4.3)
- resonance can be put to good use (e.g. wind instruments) but can cause serious damage (e.g. bridge failures)

Progressive waves

- all progressive waves travel away from their source and convey energy
- as a wave spreads out from a point source, it conveys energy over a larger area
- intensity (power per unit area) decreases with increasing distance from a point source
- as a result of energy conservation, the reduction in intensity of a wave spreading out uniformly in all directions obeys an inverse square law provided none is absorbed along the way
- waves are either mechanical (e.g. along a spring, on water, in air) or electromagnetic (e.g. visible light, microwaves)
- mechanical waves are either transverse (e.g. on water) or longitudinal (e.g. sound)
- a transverse wave consists of oscillations perpendicular to the direction of travel
- a longitudinal wave consists of oscillations parallel to the direction of travel
- all electromagnetic waves are transverse and move at 3×10^8 m s^{-1} *in vacuo*

Fig 4.3 *Response of a system that is being forced to oscillate*

- a transverse wave which has all its oscillations in a single plane is plane polarised
- only transverse waves can be polarised, longitudinal waves cannot be polarised

Wave superposition

- when the paths of waves cross, the waves pass straight through each other
- where the paths cross, the resulting displacement is the sum of the displacements of the individual waves
- displacement is a vector so waves can either reinforce (constructive superposition) or cancel (destructive superposition)
- waves in phase reinforce to produce maxima whereas those out of phase cancel to produce minima (Figure 4.4)
- complete cancellation occurs only when the out of phase waves have identical amplitudes

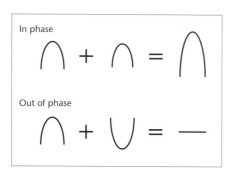

Fig 4.4 *When waves combine, their phase determines the outcome*

Two source superposition

- sources used must have the same frequency and maintain a constant phase difference (usually zero)
- such sources are said to be coherent
- waves spread out from each source and overlap as seen when two sets of circular waves are produced by two dippers in a ripple tank
- phase difference of overlapping waves is determined by how far each has travelled from its source
- for waves from two sources that are in phase:

    where path difference = $n\lambda$ the waves are still in phase and so produce a maximum
    where path difference = $(n + \frac{1}{2})\lambda$, the waves will be out of phase and so produce a minimum

- two slits can be used with a single source to produce two coherent sources
- waves spread out (diffract) as they pass through the slits
- amount of diffraction depends on the slit width and the wavelength of the waves
- the narrower the slit and the larger the wavelength, the greater the diffraction (Figure 4.5)

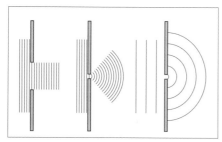

Fig 4.5 *Diffraction depends on slit width and wavelength*

Stationary waves

- produced by the superposition of two identical waves travelling in opposite directions
- resulting pattern has points where the amplitude is always zero (nodes) and those where there is a maximum amplitude (antinodes)
- adjacent nodes are half wavelength apart, as are adjacent antinodes
- all points between adjacent nodes move in phase
- points either side of a node move out of phase
- a stationary wave does not transfer energy although it does have energy 'trapped' within its antinodal regions

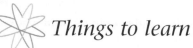 *Things to learn*

You should learn the following for your Unit PHY4 Test. Remember that it may also test your understanding of the 'general requirements' (see Appendix 1).

Equations that will *not* be given to you in the test

☐ centripetal force = mass × speed^2/radius

$F = mv^2/r$

☐ wave speed = frequency × wavelength

$v = f\lambda$ λ = wavelength

Laws

☐ inverse square law: when a quantity decreases in proportion to the square of the increasing distance

☐ principle of superposition: resultant displacement at any point is always equal to the vector sum of the displacements of the individual waves at that point

General definitions

☐ period: time to complete one revolution ... time taken to cover one complete oscillation

☐ centripetal acceleration: rate of change of velocity of a body following a circular path, directed towards the centre of the circle

☐ centripetal force: resultant force that must act towards the centre of a circle to make a body follow a circular path

☐ apparent weightlessness: situation that occurs when a body is in free fall with only the force of gravity acting on it

☐ amplitude: maximum displacement from the equilibrium position

☐ frequency: number of oscillations per second

☐ simple harmonic motion: motion where the acceleration is directly proportional to the displacement from a fixed point and always directed towards that point

☐ natural frequency: the frequency at which an isolated system oscillates after it has been displaced and then released

☐ resonance: the large-amplitude oscillations that arise as a result of a system being forced to oscillate at its natural frequency

☐ electromagnetic wave: transverse combination of oscillating electric and magnetic fields

☐ transverse wave: a wave where the oscillations are perpendicular to the direction of travel

☐ longitudinal wave: a wave where the oscillations are parallel to the direction of travel

☐ plane polarised: a transverse wave that has all its oscillations confined to a single plane perpendicular to the direction of travel

☐ coherent sources: sources that have the same frequency and maintain a constant phase difference

☐ node: a point on a stationary wave where the displacement is always zero

Unit 4

❏ antinode: a point on a stationary wave that oscillates with the maximum amplitude

Word equation definitions

Use the following word equations when asked to define:

❏ angular speed = change in central angle/time taken

Experiments

❏ 1. Oscillations of a pendulum

Set up a pendulum and measure the length l of its string (from support to centre of bob).
Displace the bob a small amount (pendulum swings with s.h.m. only when its amplitude is small).
Release the bob and time at least 10 oscillations.
Repeat, average and calculate the period T.
Vary the length to obtain a series of corresponding readings for l and T.
Plot a graph of T^2 against l.
Graph is a straight line through the origin showing that $T^2 \propto l$.
Since $T = 2\pi\sqrt{(l/g)}$
$T^2 = 4\pi^2 l/g$
so gradient of graph = $4\pi^2/g$
$g = 4\pi^2/\text{gradient}$

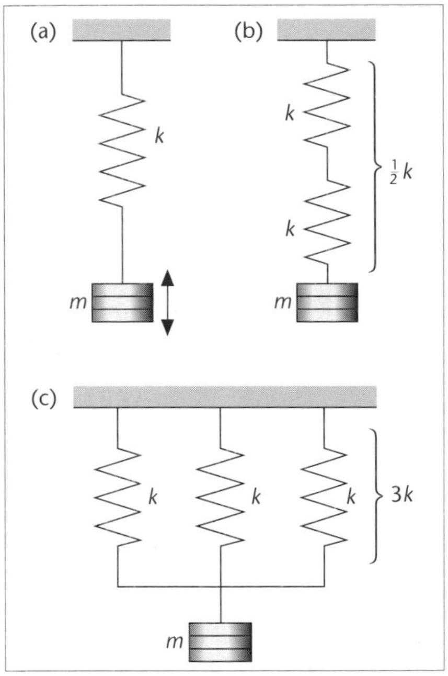

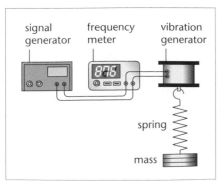

Fig 4.6 *(a) Vertical oscillations of a mass–spring system, (b) and (c) Varying the force constant*

❏ 2. Oscillations of a mass–spring system

Suspend a mass m on the end of a vertical spring (Figure 4.6a)
Displace the mass a small amount – don't stretch it beyond its elastic limit
Release the mass and time at least 10 oscillations.
Repeat, average and calculate the period T.
Vary the mass to obtain a series of corresponding readings for m and T.
Plot a graph of T^2 against m.
Graph is a straight line through the origin showing that $T^2 \propto m$.
Use different arrangements of identical springs to vary the force constant (Figures 4.6b and c)
Obtain a series of corresponding readings for k and T.
Plot a graph of T^2 against $1/k$.
Graph is a straight line through the origin showing that $T^2 \propto 1/k$.
So $T^2 \propto m/k$ and $T \propto \sqrt{(m/k)}$ which agrees with the equation $T = 2\pi\sqrt{(m/k)}$

❏ 3. Forced oscillations and resonance

Suspend a 200 g mass from a vertical spring.
Displace the mass a small amount, release and time at least 10 oscillations.
Repeat, average and calculate the period T and the natural frequency f (= $1/T$).
Attach the spring and mass to a vibration generator (Figure 4.7).
Set the signal generator to a frequency below the natural frequency.
Record the mass's maximum amplitude.
Increase the signal generator's frequency in steps to a value above the natural frequency.
Record the mass's maximum amplitude at each frequency.
Plot a graph of maximum amplitude against signal generator's frequency.
Repeat the experiment with the mass immersed in a beaker of water to increase damping.
Graphs should be similar to those in Figure 4.3.

signal generator   frequency meter   vibration generator

876

spring

mass

Fig 4.7 *The vibration generator forces the mass to oscillate*

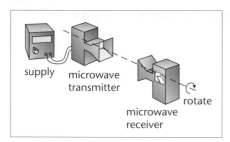

Fig 4.8 *Rotate the receiver in the direction of the arrow*

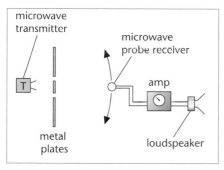

Fig 4.9 *Move the probe receiver in the directions of the arrows*

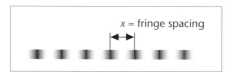

Fig 4.10 *Equally spaced fringes are seen on the screen*

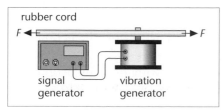

Fig 4.11 *The vibration generator forces the rubber cord to oscillate*

❏ 4. Demonstrating polarisation

Using microwaves:
Set up a microwave transmitter facing a receiver (Figure 4.8).
Rotate the receiver in the direction of the arrow.
The signal strength falls to zero and rises again twice during a full rotation.
This shows that microwaves from the transmitter are plane polarised.

Using light:
Observe some reflected light (e.g. from a shiny bench) through a Polaroid filter.
Rotate the filter.
The observed intensity falls to zero and rises again twice during a full rotation.
This shows that reflected light is plane polarised.

❏ 5. Superposition experiments

Using microwaves:
Set up the apparatus in Figure 4.9.
Use a slit width of 2 cm and a slit separation of 7 cm (use a 5 cm central plate).
Move the probe receiver as shown along an arc 30 cm from the slits.
Find the position of the first maximum intensity away from the centre.
Use a ruler to measure the distance from the centre of each slit to the probe receiver.
The difference in these two distances (path difference) is the wavelength.
Find the position of the second maximum intensity away from the centre and measure the distances.
The difference in these two distances is twice the wavelength.
Similarly, the difference in the two distances to the first minimum is half the wavelength.
Use all the values to find an average value for the wavelength.

Using light:
This experiment is often referred to as Young's double slit experiment.
Use a laser to illuminate two narrow slits with a separation s of 0.5 mm.
Observe the superposition pattern produced on a screen a distance D of 6 m from the double slits (Figure 4.10).
Use a ruler to measure the width of several fringes and calculate the fringe spacing x.
Use the equation $\lambda = xs/D$ to calculate the wavelength of the laser light.

❏ 6. Stationary waves

Using a stretched rubber cord:
Stretch a rubber cord attached to a vibration generator and clamp its ends (Figure 4.11).
Starting at a low value, slowly increase the frequency of the vibration generator.
Observe the stationary wave pattern of the fundamental frequency f.
Measure the distance between the two nodes and double it to get the wavelength λ.
Use the equation $v = f\lambda$ to calculate the speed of waves on the rubber cord.
Continue to increase the frequency slowly until the next stationary wave pattern is produced.

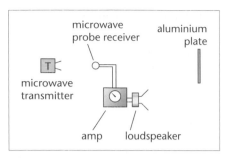

Fig 4.12 *Move the probe receiver to locate the nodes and antinodes*

The position of a node can be pinpointed more accurately than that of an antinode.

This occurs at twice the fundamental frequency.
The wavelength halves showing that the wave speed remains the same.
Repeat this procedure to observe the stationary wave patterns at 3*f*, 4*f* and 5*f*.

Using microwaves:
Point a microwave transmitter at an aluminium plate.
The aluminium reflects the microwaves, which then superpose with those leaving the transmitter.
Move a probe receiver along the line between transmitter and reflector (Figure 4.12).
Locate the positions of the nodes and antinodes.
Use a ruler to measure the total distance between 10 sets of adjacent nodes.
Divide this distance by 5 to get the wavelength of the microwaves.

 Checklist

Before attempting the following questions on oscillations and waves, check that you:

❏ know that a body moving with circular motion has both a linear speed and an angular speed

❏ know that angular speed is measured in rad s^{-1} and understand what this means

❏ can explain why a body moving at constant speed around a circular path is accelerating

❏ appreciate that its acceleration is directed towards the centre of the circle

❏ realise that a resultant force is needed towards the centre of the circle to produce this acceleration

❏ understand why this resultant force doesn't do any work

❏ know what happens to the body when the resultant force is suddenly removed

❏ understand the concept of apparent weightlessness and the condition in which this occurs

❏ can use the terms displacement, amplitude, period and frequency to describe oscillations

❏ appreciate that, as a result of frictional forces, all oscillations eventually die away

❏ have learnt a definition of simple harmonic motion

❏ appreciate why many clocks contain a simple harmonic oscillator

❏ have learnt a description of an experiment showing how the period of a pendulum depends on its length

❏ have learnt a description of an experiment showing how the period of a mass oscillating on the end of a spring depends on its mass and the force constant of the spring

❏ can sketch linked graphs showing how the displacement ($x_0 \cos \omega t$), velocity ($-\omega x_0 \sin \omega t$) and acceleration ($-\omega^2 x_0 \cos \omega t$) of a simple harmonic oscillator vary with time

❏ know that at the centre of an oscillation the velocity is maximum ($v_{max} = \omega x_0$) and the acceleration is zero

- [ ] know that at the extreme positions, the velocity is zero and the acceleration is maximum ($a_{max} = -\omega^2 x_0$)
- [ ] appreciate that all oscillators have their own natural frequency
- [ ] understand that resonance occurs when an oscillator is forced to oscillate at its natural frequency and that this results in large amplitude oscillations
- [ ] have learnt a description of an experiment to plot resonance curves for a mass oscillating on the end of a spring
- [ ] know the difference between a progressive and a stationary wave
- [ ] can use the inverse square law to calculate the intensity of a wave spreading out from a point source at various distances from the source
- [ ] know the difference between a transverse and a longitudinal wave
- [ ] appreciate that all electromagnetic waves are transverse and move at the same speed in vacuo
- [ ] have learnt the seven main regions of the electromagnetic spectrum and know the order of magnitude of their wavelength ranges (e.g. visible: 400 to 700 nm)
- [ ] understand what is meant by plane polarisation and appreciate why only transverse waves can be polarised
- [ ] have learnt a description of experiments demonstrating plane polarisation using microwaves and light
- [ ] have learnt a statement of the principle of superposition
- [ ] can explain the formation of maxima and minima in terms of the relative phase of the waves involved
- [ ] appreciate why two sources have to be coherent to form a stable superposition pattern
- [ ] understand the connection between phase difference and path difference, $\Delta(\text{phase}) = 2\pi\Delta(\text{path})/\lambda$
- [ ] know how the path difference determines whether a maximum or a minimum is formed
- [ ] appreciate the part played by diffraction in a two slit superposition experiment
- [ ] know the effects of slit width and wavelength on the amount of diffraction
- [ ] have learnt a description of experiments demonstrating two source superposition using microwaves and light and know how to find the wavelength of the waves used
- [ ] appreciate the limitations in the use of the equation $\lambda = xs/D$
- [ ] understand the conditions needed for the production of a stationary wave
- [ ] are familiar with the terms nodes and antinodes, and know how to find wavelength from the distances between them
- [ ] have learnt a description of experiments demonstrating stationary waves on a stretched rubber cord and in microwaves and know how to find the wavelength of the waves used
- [ ] are familiar with the 'general requirements' (see Appendix 1) and how they apply to the topic of oscillations and waves

Answers to these questions, together with explanations, are in the Answers section which follows Chapter 6.

Unit 4

Testing your knowledge and understanding

Quick test

Select the correct answer to each of the following questions from the four answers supplied. In each case only one of the four answers is correct. Allow about 40 minutes for the 20 questions.

1 A cogwheel having 25 teeth is rotated with an angular speed of 6π rad s^{-1}. When a thin strip of metal is positioned so that the cogs strike it, the frequency of the note heard is

    A 25 Hz   **B** 75 Hz   **C** 180 Hz   **D** 4500 Hz

2 A body travels at a constant rate along a circular path. Which of the following quantities associated with its motion does NOT remain constant?

    A Angular speed
    B Kinetic energy
    C Linear momentum
    D Linear speed

3 The speed of a satellite in a circular orbit of radius 10 Mm around the Earth is 6.3 km s^{-1}. Its acceleration is approximately

    A 0.16 m s^{-2} **B** 0.25 m s^{-2} **C** 4.0 m s^{-2} **D** 6.3 m s^{-2}

4 A spacecraft is in orbit round the Earth and is approximately 1600 km above the Earth's surface. An astronaut in the spacecraft experiences apparent weightlessness. This is because the gravitational force exerted by the Earth on the astronaut is

    A balanced by gravitational forces exerted on the astronaut by the Moon
    B completely shielded from the astronaut by the enclosing spacecraft shell
    C exactly equal to the force required to keep the astronaut moving in orbit
    D zero at this distance from the Earth

5 In a fairground shooting gallery there is a gun that automatically fires at random times at a moving target. The player has to aim the gun in one direction and leave it there. The target moves back and forth with simple harmonic motion. Five regions of the motion, labelled 1 to 5, are marked in Figure 4.13.
Which of the following indicates the region(s) at which it is sensible to take fixed aim so as to score the greatest number of hits?

    A Aim at region 1 or region 5
    B Aim at region 2 or region 4
    C Aim at region 3
    D Aim at any region as the chance is the same for all five

6 The gravitational field strength g on the Moon is one-sixth that on Earth. A pendulum with a period of 1 s on Earth would have, on the Moon, a period of

    A 1/6 s   **B** $1/\sqrt{6}$ s   **C** $\sqrt{6}$ s   **D** 6 s

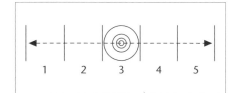

Fig 4.13

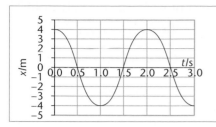

Fig 4.14

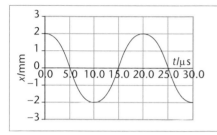

Fig 4.15

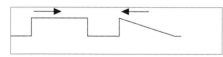

Fig 4.16

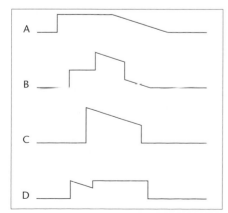

Fig 4.17

7 A particle performs simple harmonic motion, the displacement x from the equilibrium position varying with time t as shown in Figure 4.14. The maximum acceleration of the particle is

   A π m s^{-2} **B** 4π m s^{-2} **C** π^2 m s^{-2} **D** $4\pi^2$ m s^{-2}

8 A lorry driver notices that the image in the rear-view mirror is blurred when the engine runs slowly and becomes clear as the engine speed is increased. This is an example of

   A superposition
   B resonance
   C polarisation
   D coherence

9 Figure 4.15 shows how the displacement x of a particle in a progressive wave varies with time t.
Which of the following gives the values of the amplitude and frequency of the vibration of the particle?

   A Amplitude = 2 mm and frequency = 25 kHz
   B Amplitude = 2 mm and frequency = 50 kHz
   C Amplitude = 4 mm and frequency = 25 kHz
   D Amplitude = 4 mm and frequency = 50 kHz

10 Which of the following describes the type of wave produced on a stretched guitar string after it has been plucked?

   A Progressive, longitudinal, electromagnetic
   B Progressive, transverse, mechanical
   C Stationary, longitudinal, electromagnetic
   D Stationary, transverse, mechanical

11 Which of the following phenomena cannot be demonstrated using sound waves?

   A Diffraction
   B Polarisation
   C Reflection
   D Superposition

12 Yellow light (1), radiowaves (2), blue light (3), X-rays (4) and infrared (5) are all electromagnetic waves. Which of the following places them in order of increasing frequency?

   A 2, 5, 1, 3, 4
   B 2, 5, 3, 1, 4
   C 4, 1, 3, 5, 2
   D 4, 3, 1, 5, 2

13 Which of the following forms of electromagnetic waves is associated with a wave of frequency 10^{15} Hz? (Speed of electromagnetic radiation in vacuum is 3.0×10^8 m s^{-1}.)

   A Infrared
   B Ultraviolet
   C Visible light
   D X-rays

14 Figure 4.16 shows two idealised wave pulses moving towards one another on a spring.
The pulses travel at the same speed. As they pass through each other, the pulses superpose. The diagrams in Figure 4.17 show the result of this superposition at four successive instants during their interaction. Which of these idealised drawings is NOT correct?

15 Which of the following is a correct statement with regard to a two-slit experiment to produce fringes for the measurement of the wavelength of light?

    A A dark fringe occurs when the path difference between the two beams is zero

    B The distance between the slits is equal to the distance between adjacent bright (or dark) fringes

    C The overlapping beams are a result of diffraction at the two slits

    D The two slits act as incoherent sources

16 When a two-slit arrangement was set up to produce a superposition pattern on a screen using a monochromatic source of green light, the fringes were found to be too close together for accurate observation. It would be possible to increase the separation of the fringes by

    A replacing the light source with a monochromatic source of red light

    B increasing the distance between the source and the slits

    C decreasing the distance between the slits and the screen

    D increasing the distance between the two slits

17 Two sources emit coherent radiation of wavelength 3 mm. The superposition pattern at a distance of 2 m from the sources consists of fringes that are 5 cm apart. The separation of the sources is

    A 0.12 m   **B** 0.30 m   **C** 0.33 m   **D** 0.75 m

18 The following statements each make a comparison of progressive and stationary waves. Which comparison is NOT correct?

    A Energy is continually transferred along a progressive wave whereas there is no transfer of energy along a stationary wave

    B Progressive waves can be polarised whereas stationary waves cannot be polarised

    C Amplitude is either constant or gradually declines along a progressive wave whereas amplitude depends on position along a stationary wave

    D A progressive wave has no nodes or antinodes whereas a stationary wave has definite positions for nodes and antinodes

19 A taut wire is fixed at one end whilst the other end is attached to a small vibration generator. The wire is set vibrating so that there are nodes at both ends and a single node along the wire at its centre. Which of the following statements is NOT correct?

    A All points of the wire on one side of the centre vibrate in phase with each other

    B The wavelength of the waves on the wire equals the length of the wire

    C Two points on the wire at equal distances from the centre have the same amplitude

    D Any two points on either side of the centre vibrate with a phase difference of 90°

20 Two waves of equal amplitude and frequency are travelling in opposite directions along the same path with speeds of 75 m s^{-1}. Their frequency is 50 Hz. The distance between adjacent nodes of the resulting stationary wave is

    A 0.33 m   **B** 0.67 m   **C** 0.75 m   **D** 1.50 m

Worked examples

Study the following worked examples on oscillations and waves carefully. Make sure you fully understand their answers before attempting the practice assessment questions.

Worked example 1

The following statements apply to a body orbiting a planet at constant speed and at a constant height.

(i) The body is travelling at constant velocity.

(ii) The body is in equilibrium because the centripetal force is equal and opposite to the weight.

State and explain whether each statement is true or false. **[2,2]**
(Total 4 marks)

(Edexcel Module Test PH1, January 1999, Q. 1 (part))

Answer:
(i) false ✓

Velocity is a vector and its direction continually changes ✓

(ii) false ✓

Centripetal force is the resultant force producing circular motion/weight is the centripetal force ✓

Worked example 2

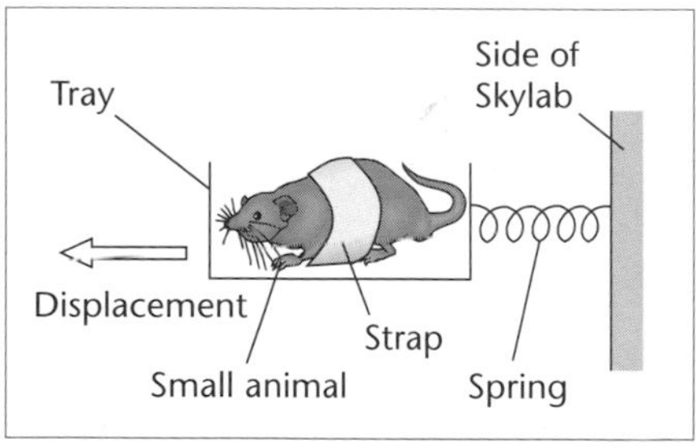

Fig 4.18

Figure 4.18 shows a method for determining the mass of small animals orbiting the Earth in a space station. The animal is securely strapped into a tray attached to the end of a spring. The tray will oscillate with simple harmonic motion when displaced as shown and released.

Define *simple harmonic motion*. **[2]**

The tray has a mass of 0.400 kg. When it contains a mass of 1.00 kg, it oscillates with a period of 1.22 s. Calculate the spring constant k. **[3]**

The 1.00 kg mass is removed and a small animal is now strapped into the tray. The new period is 1.48 s. Calculate the mass of the animal. **[3]**

The astronauts suggest that the calibration experiment with the 1.00 kg mass could have been carried out on Earth before take off. If a similar experiment were conducted on Earth would the time period be greater than, less than, or equal to 1.22 s? Explain your answer. **[3]**
(Total 11 marks)

(Edexcel Module Test PH2, June 1998, Q. 5)

Answer:

Motion where the acceleration is directly proportional to the displacement from a fixed point ✓

and always directed towards that point ✓

Total mass $m = 0.400$ kg $+ 1.00$ kg $= 1.40$ kg ✓

Using $T = 2\pi\sqrt{(m/k)}$ or $T^2 = 4\pi^2 m/k$

$k = 4\pi^2 m/T^2 = 4 \times \pi^2 \times 1.40$ kg$/(1.22$ s$)^2$ ✓

$= 37.1$ N m^{-1} (or kg s^{-2}) ✓

$m = kT^2/4\pi^2 = 37.1 \text{ N m}^{-1} \times (1.48 \text{ s})^2/(4 \times \pi^2)$ ✓

$\qquad = 2.06 \text{ kg}$ ✓

Mass of small(ish) animal = 2.06 kg − 0.400 kg = 1.66 kg ✓

The period would be equal to 1.22 s on Earth ✓
Period depends on $\sqrt{(m/k)}$ and mass is the same everywhere ✓
and k is a constant for any given spring system ✓

Worked example 3

A 60 W filament lamp transfers electrical energy to light with an efficiency of 12%. Calculate the light intensity produced by the lamp at a point 3.5 m from the filament
. **[3]**
The lamp is observed through a sheet of Polaroid. Describe and explain the effect of this on the intensity of the light. **[3]**
The sheet of Polaroid is now slowly rotated in a plane perpendicular to the direction of propagation of the light. What effect does this have on the intensity of the light?**[1]**
(Total 7 marks)
(Edexcel Module Test PH2, January 2001, Q. 5)

Answer:
Light output = 0.12 × 60 W = 7.2 W ✓
$\qquad$ Intensity = $P/4\pi r^2$ = 7.2 W/[4 × π × (3.5 m)^2] ✓
$\qquad\qquad = 0.047 \text{ W m}^{-2}$ ✓

Intensity is reduced ✓
Light from lamp is unpolarised with vibrations in many planes ✓
Polaroid removes all but one plane of vibrations ✓

Rotating the Polaroid has no effect ✓

Helpful hint

Don't fall for the trap here, the intensity will only fall and rise if the incident light is already polarised

Worked example 4

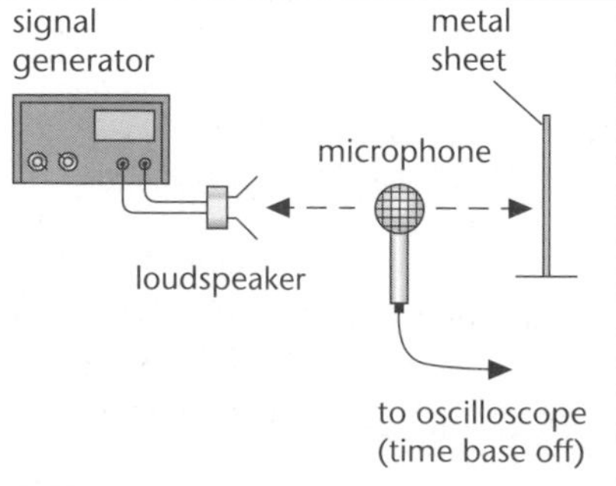

Figure 4.19 shows a loudspeaker that sends a note of constant frequency towards a vertical metal sheet.
As the microphone is moved between the loudspeaker and the metal sheet the amplitude of the vertical trace on the oscilloscope continually changes several times between maximum and minimum values. This shows that a stationary wave has been set up in the space between the loudspeaker and the metal sheet. How has the stationary wave been produced? **[3]**
State how the stationary wave pattern changes when the frequency of the signal generator is doubled. Explain your answer. **[2]**
What measurements would you take, and how would you use them, to calculate the speed of sound in air? **[4]**
Suggest why the minima detected near the sheet are much smaller than those detected near the loudspeaker. **[3]**
(Total 12 marks)
(Edexcel Module Test PH2, January 1996, Q. 4)

Answer:
Metal sheet reflects the sound waves ✓
Stationary wave is a result of the superposition ✓
between waves of the same frequency moving in opposite directions ✓

signal generator   metal sheet

microphone

loudspeaker

to oscilloscope
(time base off)

Fig 4.19

Nodes and antinodes are twice as close together ✓
since the wavelength halves when the frequency doubles ✓

Measure total distance between several nodes ✓
Calculate distance between adjacent nodes ✓
Read frequency off signal generator/use a frequency meter ✓
Wavelength = 2 × distance between adjacent nodes ✓
Speed $v = f\lambda$ ✓   **Max 4**

Near the sheet there is almost complete cancellation ✓
Incident and reflected waves have the same amplitude (as they've travelled roughly the same distance) ✓
Near loudspeaker, reflected waves have travelled a lot further so their amplitude is much less ✓

Answers to these questions, together with explanations, are in the Answers section which follows Chapter 6.

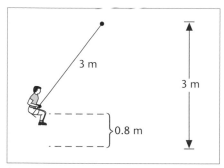

Fig 4.20

Practice questions

The following are typical assessment questions on oscillations and waves. Attempt these questions under similar conditions to those in which you will sit your actual test.

1   The period of the Earth about the Sun is approximately 365 days. Use this value to calculate the angular speed of the Earth about the Sun in rad s^{-1}. **[2]**
The mass of the Earth is 5.98×10^{24} kg and its average distance from the Sun is 1.50×10^{11} m. Calculate the centripetal force acting on the Earth. **[3]**
What provides this centripetal force? **[1]**
(Total 6 marks)
(Edexcel Module Test PH1, June 1997, Q. 8)

2   A child of mass 21 kg sits on a swing of length 3.0 m and swings through a vertical height of 0.80 m (Figure 4.20).
Calculate the speed of the child at a moment when the child is moving through the lowest position. **[3]**
Calculate the force exerted on the child by the seat of the swing at the same moment. **[4]**
(Total 7 marks)
(Edexcel Module Test PH1, January 1998, Q. 6 (part))

3   Fill in the gaps in the following sentences.
A body oscillates with simple harmonic motion when the resultant force F acting on it and its displacement x are related by the expression ....
The acceleration of such a body is always directed ....
The acceleration of the body is a maximum when its displacement is .... and its velocity is .... when its displacement is zero. **[4]**

A mass of 0.80 kg suspended from a vertical spring oscillates with a period of 1.5 s. Calculate the force (spring) constant of the spring. **[2]**
(Total 6 marks)
(Edexcel Module Test PH2, June 2000, Q. 4)

4 Calculate the period *T* of a simple pendulum of length 24.9 m. **[2]**
The pendulum is displaced by 3.25 m and allowed to swing freely.
Calculate the maximum speed of the pendulum. **[2]**
Calculate its maximum acceleration. **[2]**
Sketch two graphs showing how the velocity and the acceleration of
the pendulum vary with time. Each graph should show *two* complete
cycles and should start at the same moment in time. Add scales to
the axes of both graphs. **[4]**
(Total 10 marks)
(Edexcel Module Test PH2, January 2000, Q. 5)

5 A motorist notices that when driving along a level road at 95 km h⁻¹,
the steering wheel vibrates with an amplitude of 6.0 mm. If she
speeds up or slows down, the amplitude of the vibrations becomes
smaller. Explain why this is an example of resonance. **[3]**
Calculate the maximum acceleration of the steering wheel given that
its frequency of vibration is 2.4 Hz. **[2]**
(Total 5 marks)
(Edexcel Module Test PH2, January 2001, Q. 4)

6 (a) The minimum intensity that can be detected by a given radio
receiver is 2.2×10^{-5} W m⁻². Calculate the maximum distance
that the receiver can be from a 10 kW transmitter so that it is
just able to detect the signal. **[3]**
(b) A radio source of frequency 95 MHz is set up in front of a metal
plate. The distance from the plate is adjusted until a stationary
wave is produced in the space between them. The distance
between any node and an adjacent antinode is found to be
0.8 m. Calculate the wavelength of the wave. **[2]**
Calculate the speed of the radio wave. **[2]**
What does this suggest about the nature of radiowaves? **[1]**
(Total 8 marks)
(Edexcel Module Test PH2, January 1997, Q. 6)

7 Describe an experiment using microwaves to produce and detect a
two-slit interference pattern. You may be awarded a mark for the
clarity of your answer. **[5]**
Suggest an appropriate slit separation for this experiment. **[1]**
How could this experiment be used to obtain a value for the
wavelength of the microwaves? **[3]**
(Total 8 marks)
(Edexcel Module Test PH2, January 1999, Q. 6)

8 A laser beam of wavelength 690 nm is directed normally at parallel
slits (Figure 4.21).
Calculate the fringe spacing at the screen and sketch the pattern that
would be observed. **[4]**
This laser beam is replaced by one with a wavelength of 460 nm.
Describe how the appearance of the fringes would change. **[2]**
The two laser beams are now directed simultaneously at the slits.
Which fringes exactly overlap? **[2]**
(Total 8 marks)
(Edexcel Module Test PH2, January 2000, Q. 7)

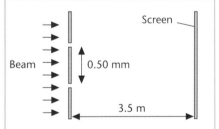

Fig 4.21

Part ❷ Quantum phenomena and the expanding universe

 Introduction

The photoelectric effect occurs when electrons are released from a metal when electromagnetic radiation with a sufficiently high frequency is shone on it. The photoelectric effect cannot be explained in terms of the wave behaviour of light. It is a quantum phenomenon and a photon model has to be used to explain it. In this section, you learn that a photon is the smallest possible packet of light energy at a given frequency. You find that all particles have the dualistic property of behaving like either a particle or a wave disturbance with the particle nature of the photon being used to explain the photoelectric effect and the wave characteristics of the photon being used to explain superposition and diffraction. You observe the wave characteristics of electrons as they pass through layers of graphite and diffract. You learn that an atom emits a photon during a transition from a higher to a lower energy state and how all the different transitions produce a characteristic spectrum. You find that such emission spectra are important in analysing the composition of stars and learn how they can be used to find the speeds at which galaxies are moving. You find that the universe is expanding and attempt to predict what will happen to it in the future.

 Things to understand

Photons

- electromagnetic radiation consists of small packets of energy (photons)
- energy of a photon depends only on the frequency of the radiation
- increasing the frequency of a beam, increases the energy of each photon ($E = hf$)
- ultraviolet photons are more energetic than infrared photons
- increasing the intensity of a beam, increases the number of photons but not their energy
- a bright blue beam contains more photons, all of the same energy, than a dim blue beam

Photoelectric effect

- energy has to be supplied to remove electrons from a metal
- surface electrons are the easiest to release
- different metals require different amounts of energy (work functions) to release their surface electrons
- a photon can release a surface electron only if it has sufficient energy

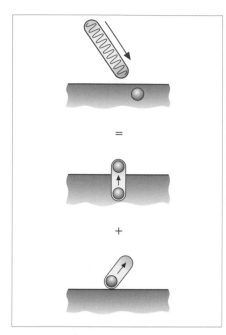

Fig 4.22 *Photon energy = work function + electron kinetic energy*

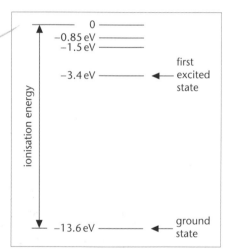

Fig 4.23 *Energy level diagram for a hydrogen atom*

- photon frequency must be above a minimum threshold value to release electrons
- if photon energy is less than the work function (frequency below threshold value) then electrons will not be released
- if photon energy is greater than the work function (frequency above threshold value), electrons are released and the excess energy is the kinetic energy of the released electron (Figure 4.22)
- the kinetic energy of a released electron can be measured by finding the potential difference needed to stop it moving (stopping potential)
- the electronvolt is a unit of energy

Emission spectra

- electrons within an atom have only a limited number of possible energies
- an energy level diagram shows the possible energies for an electron in a given atom
- usually an electron lies in the lowest energy level, known as the ground state
- a precise amount of energy is needed to move an electron up a level
- an atom with one or more electrons in raised energy levels is excited
- when an electron moves back down, the same precise amount of energy is released as a photon of frequency f where $hf = \Delta E$
- if an electron gains sufficient energy to reach the highest energy level it can leave the atom so the atom is then ionised
- an electron in the highest energy level has (by definition) zero energy and so all other electron energy levels are negative (Figure 4.23)
- free electrons only have kinetic energy and therefore their total energy is always positive
- in a gas discharge tube, atoms are continually being excited/ionised by electrons colliding with them and photons are continually released as electrons fall back down to a lower level
- the irregular spacing of the energy levels means that photons with different energies and, therefore, frequencies are released
- viewing light emitted from a gas discharge tube through a diffraction grating produces a line emission spectrum, where each line is one of the emitted discrete frequencies
- the atoms of each element have a characteristic set of energy levels and so give out a set of frequencies from which they can be identified

Wave-particle duality

- most properties of light, such as diffraction and superposition, are explained by considering it to be a wave
- a wave theory is unable to explain the photoelectric effect, particularly the existence of a threshold frequency and its independence of intensity
- a wave/photon model of light is needed to explain all of its properties
- electrons are very small, negatively charged particles
- when electrons are fired at graphite, a diffraction pattern is produced allowing the wavelength of the electrons to be calculated
- a particle/wave model of an electron is needed to explain all of its behaviour

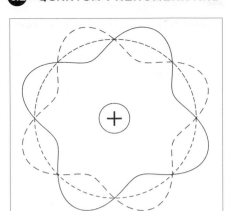

Fig 4.24 *Possible electron stationary waves in an atom*

- de Broglie's equation relates a particle's wavelength to its momentum, $\lambda = h/p$
- all particles have an associated wavelength, although for most particles the wavelength is so small that that any wave effects are unobservable
- atomic energy levels can be explained by the electrons behaving like stationary waves within the atom (Figure 4.24)
- only certain patterns of stationary wave fit and each of these corresponds to one of the fixed energy states of the atom

The expanding universe

- a star's spectrum can be produced using a diffraction grating
- dark lines across the spectrum denote certain frequencies that have been absorbed by atoms in the outer layers of the star
- these absorption lines give evidence of the elements that are present in the star
- when compared to laboratory spectra, those from distant galaxies appear to be Doppler shifted towards the red end of the spectrum
- red-shift occurs as a result of the galaxies moving away from us and indicates that the universe is expanding
- the amount of shift depends on speed and allows the recession velocities of galaxies to be measured
- the further a galaxy is from us, the greater is its recession velocity (Figure 4.25)
- dividing a galaxy's distance by its recession velocity gives the same expansion time for every galaxy
- all galaxies have been spreading out since they were formed during the Big Bang about 10^{10} years ago
- astronomical distances are often measured in light years, the distance travelled by light in one year
- the rate of expansion of the universe is probably slowly decreasing as a result of the gravitational force of attraction between the masses of the galaxies
- the fate of the universe depends on the average mass-energy density of the universe
- if this is large enough, the universe is closed and will eventually stop expanding and begin to collapse inwards, producing a Big Crunch
- if the average mass-energy density is insufficient, the universe is open and will go on expanding for ever but at a slower and slower rate

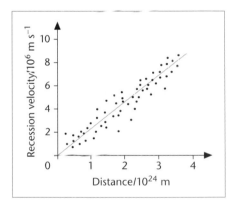

Fig 4.25 *Despite the uncertainty in measuring astronomical distances, the data suggests that $v = Hd$*

 Things to learn

You should learn the following for your Unit PHY4 Test. Remember that it may also test your understanding of the 'general requirements' (see Appendix 1).

Equations that will *not* be given to you in the test

☐ wave speed = frequency × wavelength

$v = f\lambda$ $\qquad\qquad\qquad \lambda =$ wavelength

Laws

- ❏ Hubble's law: recession velocity of galaxy ∝ distance of galaxy from us

    $$v = Hd$$ where H is the Hubble constant

General definitions

- ❏ threshold frequency: minimum frequency of electromagnetic radiation that will cause photoelectric emission from the surface of a metal
- ❏ threshold wavelength: maximum wavelength of electromagnetic radiation that will cause photoelectric emission from the surface of a metal
- ❏ work function: amount of energy needed to release an electron from the surface of a metal
- ❏ electronvolt: energy transferred to an electron charge when it moves through a potential difference of 1 V
- ❏ excitation energy: minimum energy needed to raise an electron within an atom to a position above its lowest energy state
- ❏ ionisation energy: minimum energy needed to free an electron from an atom, leaving behind a positive ion
- ❏ Doppler effect: apparent change in the frequency of light caused by relative movement between the source and the observer
- ❏ red-shift: an apparent decrease in frequency caused by the source and the observer moving apart
- ❏ closed universe: the average mass-energy density of the universe is sufficient for gravitational forces eventually to pull it back together
- ❏ open universe: the average mass-energy density is insufficient and the universe will continue to expand, although at a decreasing rate

Experiments

❏ 1. Simple demonstration of the photoelectric effect

Remove the oxide layer from the surface of a zinc plate.
Put the zinc plate on the cap of a gold-leaf electroscope.
Give the electroscope a negative charge so that the gold leaf deflects.
Observe what happens to the gold leaf when first red and then ultraviolet light is shone onto the zinc plate (Figure 4.26).
The gold leaf remains deflected when red light is used.
The gold leaf gradually falls back down when ultraviolet light is used.
If the electroscope is charged positively, the gold leaf remains deflected whatever radiation is used.

❏ 2. Measuring the Planck constant

Connect a photocell to the circuit shown in Figure 4.27.
Set the potentiometer so that the voltmeter reads zero.
Shine light of known frequency onto the emitting electrode and observe the current flowing in the picoammeter.
Slowly increase the potential difference across the cell until the current decreases to zero.
Record the stopping potential from the voltmeter.
Repeat for a range of frequencies.
Plot a graph of stopping voltage against frequency (Figure 4.28).

Helpful hint

1 eV = 1.6×10^{-19} J is given in the data at the end of your test paper (see Appendix 2)

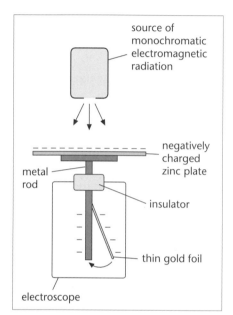

Fig 4.26 *The deflection of the gold leaf is a measure of the charge stored on the zinc plate*

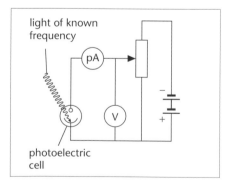

Fig 4.27 *Measuring the energy of the photoelectrons*

Helpful hint

This apparatus can be used to show how photocurrent varies with potential difference across the photocell

---

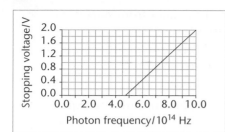

Fig 4.28 *Stopping potential for photoelectrons from a caesium electrode for a range of frequencies*

Photon energy = work function + maximum kinetic energy

= work function + (electronic charge × stopping potential)

$hf = \varphi + eV_s$

$V_s = (h/e) \times f - \varphi/e$

Comparing this with $y = mx + c$

Gradient = h/e

So $h = e \times$ gradient

 Checklist

Before attempting the following questions on quantum phenomena and the expanding universe, check that you:

❑ know that all electromagnetic radiation consists of small packets of energy called photons

❑ can use the equations $E = hf$ and $E = hc/\lambda$ to calculate the energy of a photon

❑ understand that an intense beam has more photons than a dim beam

❑ appreciate that if the photon energy is less than the work function, no photoelectrons will be released whatever the intensity of the incident beam

❑ know that even a dim beam will release photoelectrons when its photon energy is more than the work function

❑ have learnt a description of an experiment to demonstrate the photoelectric effect using a zinc plate and a golf-leaf electroscope

❑ can calculate the maximum speed of photoelectrons emitted from a given surface by a known frequency (or wavelength) of incident radiation

❑ know how to convert between electronvolts and joules

❑ appreciate how stopping potential is used to measure the maximum kinetic energy of photoelectrons

❑ have learnt a description of an experiment to measure the Planck constant

❑ know that an atom has a limited number of fixed energy states and that these are best displayed on an energy level diagram

❑ appreciate why all energy levels are negative

❑ understand the difference between excitation and ionisation energies

❑ can calculate the frequency of the photon emitted when an atom moves to a lower energy state

❑ know how a diffraction grating can be used to observe line emission spectra

❑ appreciate that each element has its own characteristic emission spectrum

❑ understand the need for a wave-particle model

❑ know that fast-moving electrons behave like waves and produce a diffraction pattern after passing through layers of graphite

❑ know that the fixed energy states of an atom can be explained in terms of stationary electron waves

❑ can calculate the de Broglie wavelength of a moving particle

Unit 4

❑ appreciate how the absorption spectra of the light from stars gives information about their chemical compositions

❑ know that relative movement between source and observer produces an apparent change in the frequency of light received

❑ can calculate speed from the amount of Doppler shift

❑ understand that red shift of light from galaxies suggests the expansion of the universe

❑ have learnt a statement of Hubble's law

❑ can use the Hubble constant to calculate the approximate age of the universe

❑ appreciate the reasons for this being only an approximate value

❑ know that the expanding universe results from a Big Bang origin

❑ understand how the fate of the universe will be determined by its average mass-energy density and appreciate that the actual value of this is unknown

❑ are familiar with the 'general requirements' (see Appendix 1) and how they apply to the topic of quantum phenomena and the expanding universe

Testing your knowledge and understanding

Quick test

Answers to these questions, together with explanations, are in the Answers section which follows Chapter 6.

Select the correct answer to each of the following questions from the four answers supplied. In each case only one of the four answers is correct. Allow about 30 minutes for the 15 questions.

1 The unit for the Planck constant h may be written as

    A $J\,s^{-1}$ **B** $N\,m\,s$ **C** $kg\,m^2\,s$ **D** $N\,m\,s^{-1}$

2 To which of the following forms of electromagnetic radiation does a photon with energy 3×10^{-19} J belong?

    A Microwaves
    B Infrared
    C Visible light
    D Ultraviolet

3 A laser emits monochromatic light of wavelength λ at a constant power P. If h is the Planck constant and c is the speed of light in a vacuum, the number of photons emitted per second by the laser is given by

    A $Pc/h\lambda$ **B** $\lambda c/Ph$ **C** $P/hc\lambda$ **D** $P\lambda/hc$

4 An ultraviolet light source causes the emission of photoelectrons from a zinc plate. A more intense light source of the same wavelength would give

    A the same number of photoelectrons each second with the same maximum energy

    B the same number of photoelectrons each second with a greater maximum energy

    C more photoelectrons each second with the same maximum energy

    D more photoelectrons each second with a greater maximum energy

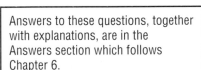

5 The maximum speed of photoelectrons emitted from a metal surface by electromagnetic radiation depends upon

    A the frequency of the radiation only

    B the intensity and the frequency of the radiation

    C the intensity of the radiation and the work function of the metal

    D the frequency of the radiation and the work function of the metal

6 Which of the following statements best describes the work function of a metal?

    A A measure of the maximum kinetic energy of photoelectrons

    B The energy of any photon that causes emission of a photoelectron

    C A measure of the minimum kinetic energy of photoelectrons

    D The minimum energy required to remove a photoelectron from the metal

7 The threshold frequency for a certain metal surface is 5.0×10^{14} Hz. The surface is illuminated with light of frequency 7.0×10^{14} Hz. The maximum kinetic energy of photoelectrons released from the surface is

    A 1.3×10^{-19} J   **B** 3.2×10^{-19} J   **C** 4.6×10^{-19} J

    D 4.6×10^{49} J

8 Which of the following is evidence for the existence of discrete electron energy levels in atoms?

    A The spectrum of a tungsten filament lamp

    B The spectrum of a mercury discharge lamp

    C The photoelectric effect

    D The Doppler effect

9 When an electron falls from an energy level of energy E_1 to one of energy E_2, radiation of frequency f and wavelength λ is emitted. If h is the Planck constant and c is the speed of light, $E_1 - E_2$ is equal to

    A hf **B** $h\lambda$ **C** hf/c **D** $h\lambda/c$

10 The ionisation energy of hydrogen is 2.2×10^{-18} J. The minimum wavelength of radiation emitted when a proton combines with an electron to form a hydrogen atom is

    A 4.8×10^{-60} m   **B** 9.0×10^{-8} m   **C** 1.1×10^{7} m

    D 1.0×10^{24} m

11 Figure 4.29 shows five energy levels of an atom. Five transitions between the energy levels are indicated, each of which will produce a photon of definite energy and frequency. Which of the four spectra shown best corresponds to these transitions?

12 The de Broglie wavelength of an electron moving at 3.0×10^{7} m s^{-1} is

    A 2.7×10^{-23} m   **B** 2.4×10^{-11} m   **C** 4.1×10^{10} m

    D 3.7×10^{22} m

13 In which of the following situations, would the Doppler effect not be observed?

    A Source and observer moving with the same velocity

    B Source and observer moving with opposite velocities

    C Source moving towards a stationary observer

    D Observer moving towards a stationary source

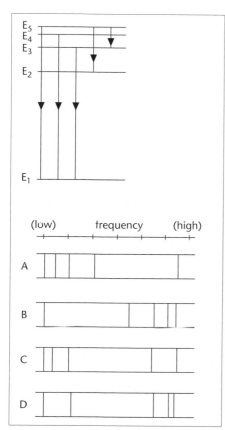

Fig 4.29

Unit 4

14 When observed from the Earth, the light from a particular galaxy contains a wavelength of 6.105×10^{-7} m. The same spectral line when measured from a laboratory source has a wavelength of 5.893×10^{-7} m. The reason for this apparent change in wavelength is that the galaxy is

    A   moving away from the Earth at 1.04×10^7 m s^{-1}
    B   moving towards the Earth at 1.04×10^7 m s^{-1}
    C   moving away from the Earth at 1.08×10^7 m s^{-1}
    D   moving towards the Earth at 1.08×10^7 m s^{-1}

15 Which of the following statements about the possible fates of the universe is not correct?

    A   If the average mass-energy density is above a critical value then the universe is closed
    B   If the universe is open then it will continue to expand but at a decreasing rate
    C   If the universe is closed then it will continue to expand but at an increasing rate
    D   If the average mass-energy density is below a critical value then the universe is open

Worked examples

Study the following worked examples on quantum phenomena carefully. Make sure you fully understand their answers before attempting the practice assessment questions.

Worked example 1

Ultraviolet light of wavelength 12.2 nm is shone on to a metal surface. The work function of the metal is 6.20 eV. Calculate the maximum kinetic energy of the emitted photoelectrons.   **[4]**
Show that the maximum speed of these photoelectrons is approximately 6×10^6 m s^{-1}.   **[2]**
Calculate the de Broglie wavelength of photoelectrons with this speed.   **[2]**
Explain why these photoelectrons would be suitable for studying the crystal structure of a molecular compound.   **[2]**
    (Total 10 marks)
(Edexcel Module Test PH2, January 2001, Q. 3)

Answer:
Work function $= 6.20$ eV $\times 1.6 \times 10^{-19}$ J eV$^{-1} = 9.92 \times 10^{-19}$ J ✓
photon energy $= hf = hc/\lambda = 6.63 \times 10^{-34}$ J s $\times 3.00 \times 10^8$ m s$^{-1}/(12.2 \times 10^{-9}$ m) ✓
        $= 1.63 \times 10^{-17}$ J ✓

Kinetic energy $= 1.63 \times 10^{-17}$ J $- 9.92 \times 10^{-19}$ J $= 1.53 \times 10^{-17}$ J ✓

$\frac{1}{2}mv^2 = \frac{1}{2} \times 9.11 \times 10^{-31}$ kg $\times v^2 = 1.53 \times 10^{-17}$ J ✓
$v^2 = 2 \times 1.53 \times 10^{-17}$ J$/(9.11 \times 10^{-31}$ kg)
$v = \sqrt{(3.36 \times 10^{13}}$ m^2 s$^{-2}) = 5.8 \times 10^6$ m s^{-1} ✓

Momentum $p = mv = 9.11 \times 10^{-31}$ kg $\times 5.8 \times 10^6$ m s$^{-1} = 5.3 \times 10^{-24}$ kg m s^{-1} ✓
$\lambda = h/p = 6.63 \times 10^{-34}$ J s$/(5.3 \times 10^{-24}$ kg m s$^{-1}) = 1.26 \times 10^{-10}$ m ✓

Diffraction would occur ✓
as wavelength is similar to the spacing/size of atoms/molecules ✓

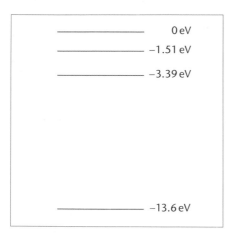

Fig 4.30

Worked example

Figure 4.30 shows some of the energy levels for atomic hydrogen.
Add arrows to the diagram showing all the single transitions that could ionise the atom. **[2]**
Why is the level labelled −13.6 eV called the ground state? **[1]**
Identify the transition that would result in the emission of light of wavelength 660 nm. **[4]**

(Total 7 marks)
(*Edexcel Module Test PH2, June 2000, Q. 9*)

Answer:
Arrows added to diagram showing:
a transition from − 13.6 eV up to 0 eV ✓
transitions from − 3.39 eV up to 0 eV **and** from −1.51 eV up to 0 eV ✓

Ground state: lowest energy state of the atom ✓

Photon energy = hc/λ = 6.63 × 10^{-34} J s × 3.00 × 10^8 m s^{-1}/(660 × 10^{-9} m) ✓
= 3.01 × 10^{-19} J/(1.6 × 10^{-19} J eV^{-1}) ✓
= 1.88 eV ✓
so need a downward transition between levels with an energy difference of 1.88 eV
transition is from −1.51 eV down to −3.39 eV ✓

Answers to these questions, together with explanations, are in the Answers section which follows Chapter 6.

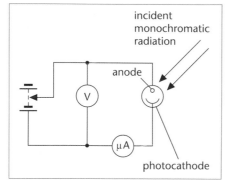

Fig 4.31

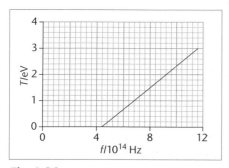

Fig 4.32

Practice questions

The following are typical assessment questions on quantum phenomena and the expanding universe. Attempt these questions under similar conditions to those in which you will sit your actual test.

1 Figure 4.31 shows monochromatic light falling on a photocell.

As the reverse potential difference between the anode and cathode is increased, the current measured by the micro-ammeter decreases. When the potential difference reaches a value V_S, called the stopping potential, the current is zero. Explain these observations. **[5]**
What would be the effect on the stopping potential of

(i) increasing only the intensity of the incident radiation,
(ii) increasing only the frequency of the incident radiation? **[2]**

(Total 7 marks)
(*Edexcel Module Test PH2, June 1998, Q. 7*)

2 The graph in Figure 4.32 shows how the maximum kinetic energy T of photoelectrons emitted from the surface of sodium metal varies with the frequency f of the incident radiation.

Why are no photoelectrons emitted at frequencies below 4.4 × 10^{14} Hz? **[1]**
Calculate the work function ϕ of sodium in electronvolts. **[3]**
Explain how the graph supports the photoelectric equation $hf = T + \phi$ **[2]**
How could the graph be used to find a value for the Planck constant? **[1]**

Unit 4

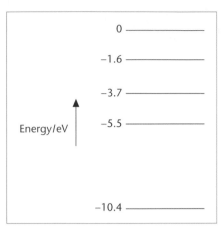

Fig 4.33

Add a line to the graph to show the maximum kinetic energy of the photoelectrons emitted from a metal which has a greater work function than sodium.   **[2]**

(Total 9 marks)

(Edexcel Module Test PH2, January 1996, Q. 8)

3   Figure 4.33 shows some of the outer energy levels of the mercury atom.
Calculate the ionisation energy in joules for an electron in the −10.4 eV level.   **[2]**
An electron has been excited to the −1.6 eV energy level. Show on the energy level diagram all the possible ways it can return to the −10.4 eV level.   **[3]**
Which change in energy levels will give rise to a yellowish line ($\lambda \approx 600$ nm) in the mercury spectrum?   **[4]**

(Total 9 marks)

(Edexcel Module Test PH2, June 1996, Q. 8)

4   In a nuclear reaction, neutrons are emitted each with a kinetic energy of 8.0×10^{-21} J. The mass of a neutron is 1.0087 u. Calculate the momentum of one of these neutrons.   **[4]**
Show that the de Broglie wavelength of these neutrons is approximately 10^{-10} m.   **[1]**
Would neutrons of this de Broglie wavelength be suitable for diffraction studies of molecular structure? Explain your answer.   **[2]**

(Total 7 marks)

(Edexcel Module Test PH2, January 1999, Q. 8)

5   An astronomical body is moving relative to the Earth. Information about its velocity may be obtained by measuring its Doppler shift. Outline the principle of this method, and suggest one of its limitations as a means of determining velocity.   **[3]**
A certain galaxy G, visible in the constellation Ursa Major, is thought to be moving away from the Earth at a speed of 1.5×10^{7} m s^{-1}. Calculate the apparent wavelength, measured using light from this galaxy, of a spectral line whose normal wavelength is 396.8 nm.   **[3]**
The Hubble constant H has a value of approximately 1.7×10^{-18} s^{-1}. Estimate the distance of the galaxy G from the Earth.   **[2]**
Use the information about galaxy G to estimate how long, in years, it has taken to reach its present distance from Earth.
(1 year = 3.2×10^{7} s)   **[3]**
Give one reason (other than the approximate value of H) why your answer may not be a reliable estimate of the time which has passed since the Big Bang.   **[1]**

(Total 12 marks)

(Edexcel Module Test PH2, June 1999, Q. 8(a))

5 Fields and forces and the A2 practical test

Part ① Fields and forces

 ### Introduction

Fields are used to interpret interactions between particles, a field being a region within which a force can be experienced. Gravitational, electric and magnetic interactions all have fields associated with them. As you study fields and forces, you learn about uniform and radial fields and the similarities and differences between gravitational and electric fields. You find that radial gravitational forces are responsible for the orbit of the planets around the Sun and learn how to calculate the period of an artificial satellite in orbit around the Earth. You find that uniform electric forces accelerate electrons towards the screen of your television and learn how to calculate the electrons' speed. You discover that a 'pulse' of current flows in a circuit in which there is a capacitor, a component containing an insulator, and find that capacitors can store energy. You investigate the magnetic fields resulting from permanent magnets and from currents flowing in wires, and discover how, when these are combined, they produce a force. You learn that whenever there is relative movement between a conductor and a magnetic field, an electromotive force is produced and how this is used to generate and to transform electricity.

Things to understand

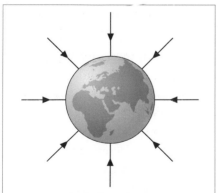

Fig 5.1 *The Earth's field is radial and directed towards the centre of the Earth*

Gravitational fields

- mass is the amount of matter in a body and, for a given body, remains the same everywhere
- all masses attract, exerting equal and opposite gravitational forces on each other
- gravitational forces are very small unless at least one of the masses involved is extremely large
- the region surrounding a mass in which another mass experiences a gravitational force of attraction is called a gravitational field
- field lines are used to show the shape and direction of a gravitational field (Figure 5.1)
- the strength of a radial gravitational field weakens with increasing

Unit 5

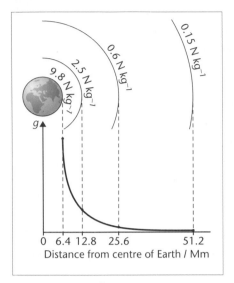

Fig 5.2 *$g \propto 1/r^2$; when distance is ×8, g is ÷64*

distance from the mass, indicated by the field lines spreading out in Figure 5.1

- for a radial field, gravitational field strength obeys an inverse square law (Figure 5.2)
- across a relatively small region, the Earth's gravitational field is effectively uniform
- the strength of a uniform gravitational field is the same everywhere
- gravitational field strength is the gravitational force that would be exerted on a mass of 1 kg placed at that point
- gravitational field strength, a vector quantity, is numerically the same as the free fall acceleration at that point, $N\ kg^{-1} = m\ s^{-2}$
- orbiting satellites have only the force of gravity acting on them
- their orbital speed is such that the gravitational force provides the exact amount of centripetal force required for their circular motion
- work is done moving masses apart resulting in an increase in their potential energy
- an orbiting satellite has constant potential energy and follows a line along a spherical equipotential surface

Electric fields

- rubbing a polythene strip with a duster charges the strip negatively and the duster positively
- electrons are dragged from the duster onto the strip leaving the strip with excess electrons (negative) and the duster with excess protons (positive)
- both the strip and the duster are insulators and so maintain their charges – a metal rod would immediately conduct away any charge transferred to it
- electric charges either attract when signs are opposite or repel when signs are the same
- the region surrounding a charge in which another charge experiences an electric force is called an electric field
- field lines are used to show the shape and direction of an electric field
- the electric field of a point charge is radial and directed towards a negative charge and away from a positive charge (Figure 5.3)
- in the central region between two parallel, oppositely charged plates, the electric field is uniform and directed towards the negative plate
- the strength of a radial electric field weakens with increasing distance from the charge whereas that of a uniform electric field is the same everywhere
- electric field strength, a vector quantity, is the electric force that would be exerted on a charge of 1 C placed at that point, measured in $N\ C^{-1}$
- for a radial field, electric field strength obeys an inverse square law
- the electric field strength of a uniform field can be found by dividing the potential difference across the plates by their separation, $N\ C^{-1} = V\ m^{-1}$
- uniform electric fields are used to accelerate electron beams
- work done, $e\Delta V$, on the electron appears as kinetic energy of the electron

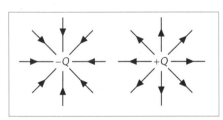

Fig 5.3 *The arrows show the directions of the forces that would act on a positive charge*

Capacitors

- a capacitor consists of two metal plates separated from each other by an insulating material

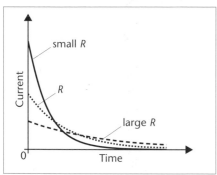

Fig 5.4 *Charging a capacitor to the same potential difference through three different resistances – the area under all three graphs, the charge displaced, is the same*

● current cannot flow in the insulator but a battery can take electrons from one plate and add electrons to the other

● one plate of a charged capacitor is positive (electrons removed) and the other is negative (electrons added)

● the amount of charge displaced depends on the potential difference across the capacitor's plates, $Q = CV$

● as a capacitor charges, the current charging it decreases

● the area under a current–time graph represents the charge displaced

● initial current is determined by the supply voltage and the total circuit resistance

● decreasing resistance increases the charging current and the capacitor takes less time to charge (Figure 5.4)

● connecting capacitors in parallel increases the total capacitance

● the total capacitance of capacitors in series is always less than any of the individual capacitances

● a charged capacitor stores energy

● the area under a potential difference–charge graph represents the energy stored

Magnetic fields

● two permanent magnets either attract when poles are opposite or repel when poles are the same

● a magnetic field is any region within which magnetic forces act

● field lines are used to show the shape and direction of a magnetic field

● all magnetic field lines are continuous

● the arrows show the direction in which the north end of the needle of a plotting compass would point

● neutral points are formed where magnetic fields overlap and completely cancel

● a current flowing in a wire produces a circular magnetic field around the wire

● the magnetic field outside a current-carrying solenoid resembles that of a bar magnet, whereas that inside it is uniform

● a force acts when a current flows at an angle to a magnetic field, the force being greatest when the current and the magnetic field are perpendicular

● the force is perpendicular to the current and the magnetic field (Figure 5.5)

● the strength of a magnetic field is called its magnetic flux density

● magnetic flux density is a vector quantity that can be measured using a pre-calibrated Hall probe

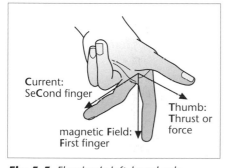

Current: SeCond finger

Thumb: Thrust or force

magnetic Field: First finger

Fig 5.5 *Fleming's left-hand rule*

Electromagnetic induction

● magnetic flux is the product of the magnetic flux density and the area through which it is passing

● the magnetic flux linkage through a coil depends on the number of turns in the coil

● an e.m.f. is induced in a coil whenever there is a change in magnetic flux linkage

Unit 5

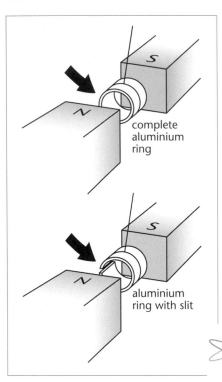

Fig 5.6 *The complete ring is the first to stop swinging as the current induced in it opposes its motion*

- the faster the rate of change in magnetic flux linkage, the larger the induced e.m.f., ε = rate of change of magnetic flux linkage
- if there is a complete circuit then an induced current flows
- induced currents flow in such a way that their interaction with the magnetic field produces an effect that opposes whatever caused them (Figure 5.6)
- a transformer uses electromagnetic induction to change the size of an alternating voltage
- alternating current in the primary coil produces an alternating magnetic field in the transformer's core
- the alternating magnetic flux through the secondary coil leads to an induced e.m.f. across its terminals
- a step-down transformer has fewer secondary than primary turns so its output voltage is less than the input
- an ideal transformer would be 100% efficient and its output power would equal its input power

 Things to learn

You should learn the following for your Unit PHY5 Test. Remember that it may also test your understanding of the 'general requirements' (See Appendix 1).

Equations that will *not* be given to you in the test

☐ weight = mass × gravitational field strength

weight = mg

☐ the inverse square law for force between two point masses

$F = Gm_1 m_2/r^2$

☐ centripetal force = mass × speed^2/radius

$F = mv^2/r$

☐ the inverse square law for force between two point charges

$F = kQ_1 Q_2/r^2$

☐ capacitance = charge displaced/potential difference

$C = Q/V$

☐ the relationship between the potential difference across the coils of a transformer and the number of turns in them

$V_1/V_2 = N_1/N_2$

Laws

☐ Newton's law of gravitation: the gravitational force between two bodies is directly proportional to the product of their masses and inversely proportional to the square of their separation

☐ Coulomb's law: the electric force between two charges is directly proportional to the product of their charges and inversely proportional to the square of their separation

☐ Faraday's law: the magnitude of the e.m.f. induced across a circuit is equal to the rate of change of magnetic flux linkage through that circuit

☐ Lenz's law: any current driven by an induced e.m.f. opposes the change causing it

General definitions

☐ gravitational field: region in which a gravitational force acts on a mass

☐ gravitational field strength: force exerted per unit mass by a gravitational field

☐ geostationary satellite: orbits above equator, in same direction as Earth rotates, with a period of 24 h and so maintains the same position above the Earth's surface

☐ inverse square law: when a quantity decreases in proportion to the square of the distance

☐ electric field: region in which an electric force acts on a charge

☐ electric field strength: force exerted per unit charge by an electric field

☐ equipotential surface: surface over which the potential is equal at all points

☐ magnetic field: region in which a magnetic force acts

☐ neutral point: position within overlapping magnetic fields where the resultant field is zero

☐ tesla: the unit of magnetic flux density; the magnetic flux density that produces a force of 1 N on each metre length of wire carrying a current of 1 A perpendicular to the magnetic field

Word equation definitions

☐ capacitance = charge displaced/potential difference   so $1\text{ F} = 1\text{ C V}^{-1}$

☐ magnetic flux density = force/(current × length of wire)

☐ magnetic flux = magnetic flux density × area

☐ magnetic flux linkage = magnetic flux × number of turns

Experiments

☐ 1. Current in a capacitor circuit

Set up the circuit shown in Figure 5.7.

With the capacitor shorted out, record the current from the micro-ammeter.

Remove the shorting lead and start timing.

Record the current every 5 s until it has fallen to 5% of its starting value. Plot a graph of current against time.

Repeat using resistances of 50 and 200 kΩ (see Figure 5.4).

Find the area under each curve either by counting squares or by approximating it to a suitable triangle.

Divide the area (charge) by the voltmeter reading to obtain a value for the capacitance.

☐ 2. Charging a capacitor at a constant rate

This experiment overcomes the problem of finding the area under a curve by keeping the charging current constant and so changes the curve into a straight line.

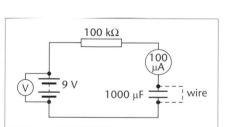

Fig 5.7 *Measuring the current charging a capacitor*

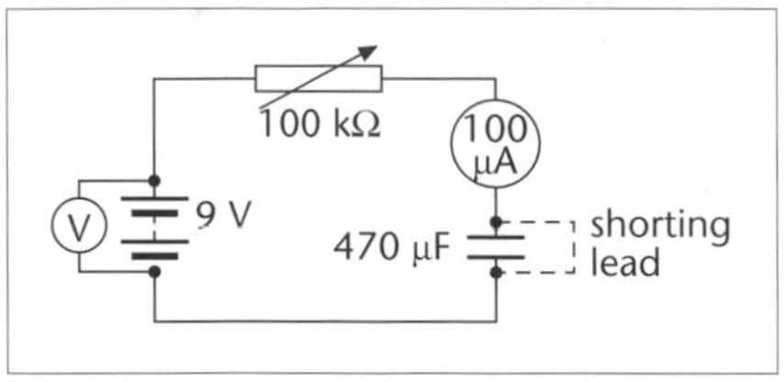

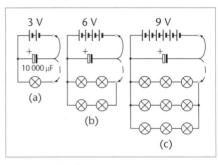

Fig 5.8 *As the capacitor charges, the resistance is decreased to keep constant current*

Helpful hint

All three circuits in Figure 5.9 have the same total resistance

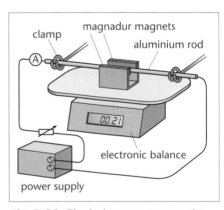

Fig 5.9 *Discharging a capacitor through lamps*

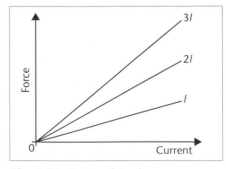

Fig 5.10 *The balance measures the force exerted on the yoke*

With the variable resistor at its maximum value, set up the circuit shown in Figure 5.8.

Short out the capacitor and adjust the variable resistor so the ammeter reads a convenient value, e.g. 80 μA

Remove the shorting lead and start timing.

Keep adjusting the variable resistor to maintain a constant current.

Stop timing when the variable resistor reaches its minimum value.

Repeat to obtain an average charging time.

Calculate charge = current × time.

Divide the charge by the voltmeter reading to obtain a value for the capacitance.

Repeat for different values of constant current.

☐ 3. Energy stored in a charged capacitor

Charge a capacitor to a potential difference of 3 V and discharge it through a single lamp (Figure 5.9a).

Note the brightness and length of flash.

Charge the capacitor to 6 V and discharge it through different lamp combinations.

Brightness and length of flash is the same when using four lamps connected as shown in Figure 5.9b.

When voltage doubles, flash energy quadruples.

Charge the capacitor to 9 V and discharge it through different lamp combinations.

Brightness and length of flash is the same when using nine lamps connected as shown in Figure 5.9c.

When voltage trebles, flash energy is nine times greater.

So energy stored in a charged capacitor $\propto V^2$.

☐ 4. Force on a current in a magnetic field

Set up the apparatus shown in Figure 5.10 with the aluminium rod clamped firmly in place.

Zero the balance, set to read force, to allow for the weight of the magnets and yoke.

Switch on the supply and adjust for a current of 1 A.

Record the force from the balance.

Vary the current to obtain a series of corresponding values of force and current.

Plot a graph of force against current.

Repeat for different lengths of rod in the magnetic field by adding further yokes of magnadur (Figure 5.11).

Increase the magnetic field strength by holding the magnadur magnets closer together and observe that the force increases.

Fig 5.11 *F ∝ l and F ∝ I*

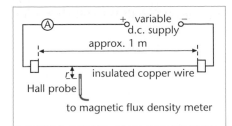

Fig 5.12 *The pre-calibrated Hall probe measures the magnetic flux density*

Helpful hint

A pre-calibrated Hall probe can be used in a similar way to investigate the magnetic field inside and outside a current-carrying solenoid

❑ 5. Investigating the magnetic field near a straight wire

Place a pre-calibrated Hall probe 1 cm from a straight wire carrying a current of 2 A (Figure 5.12).

Observe the reading on the flux density meter as the probe is slowly rotated.

Record the maximum reading; probe is then at 90° to the field lines.

Repeat for currents increasing in steps of 2 A, up to a maximum of 10 A.

Plot a graph of magnetic flux density against current.

Straight line through the origin shows that $B \propto I$.

Using a current of 10 A, record the magnetic flux density at different distances from the wire.

Plot a graph of magnetic flux density against 1/distance.

Straight line through the origin shows that $B \propto 1/r$.

 Checklist

Before attempting the following questions on fields and forces, check that you:

❑ appreciate that all masses attract

❑ understand the concept of a field and how this applies to masses, charges and magnets/currents

❑ can sketch the shape and show the direction of the radial gravitational field of a mass

❑ have learnt a statement of Newton's law of gravitation

❑ appreciate what is meant by an inverse square law

❑ know why the gravitational field strength is numerically the same as the free fall acceleration

❑ know that gravitational forces are responsible for the circular motion of planets and satellites

❑ appreciate the usefulness of geostationary satellites

❑ understand that the potential energy of a satellite remains constant

❑ realise that equipotential lines are always at 90° to field lines

❑ know that under certain conditions the Earth's gravitational field can be considered as uniform

❑ understand the process whereby an object is charged by frictional contact

❑ know that there are two signs of charge and that the electronic charge is the smallest possible amount of charge

❑ appreciate that opposite charges attract and like charges repel

❑ can sketch the radial electric field of both positive and negative charges

❑ have learnt a statement of Coulomb's law

❑ can sketch the electric field between two, oppositely charged parallel plates and can add a number of equipotentials to it

❑ know how a uniform electric field is used to accelerate charged particles

❑ can calculate the speed of an electron fired from an electron gun

❑ know the structure of a capacitor and why it cannot conduct a current

❑ understand what is happening when a capacitor is charging

❑ have learnt a description of an experiment to show how the current charging a capacitor varies with time and know how a variable resistor can be used to maintain a constant charging current

❏ know that the area under a current–time graph represents charge

❏ have learnt the definition of capacitance and know its unit

❏ understand the effect of combining capacitors in series and in parallel

❏ know that a charged capacitor stores energy

❏ have learnt a description of an experiment to show how the energy stored in a charge capacitor depends on the potential difference across it

❏ can sketch the magnetic field produce by a single bar magnet, between two attracting magnets and between two repelling magnets

❏ understand the concept of a neutral point

❏ can sketch the magnetic fields produced by a current in a wire and in a solenoid

❏ know that a force acts on a current flowing at an angle to a magnetic field

❏ can use Fleming's left-hand rule to find the direction of this force

❏ have learnt a description of an experiment to investigate the force exerted on a current in a magnetic field

❏ have learnt the definition of magnetic flux density and its unit, the tesla

❏ appreciate that magnetic flux density can be measured using a Hall probe

❏ have learnt a description of an experiment to investigate the magnetic field near a straight wire and know how a similar method can be used for a solenoid

❏ can distinguish between magnetic flux density, magnetic flux and magnetic flux linkage

❏ know that any change in magnetic flux through a coil induces an e.m.f.

❏ have learnt a statement of Faraday's law

❏ have learnt a statement of Lenz's law

❏ understand the principle of operation of a transformer

❏ know that a transformer will not work using direct current

❏ can calculate the output voltage from a known transformer

❏ are familiar with the 'general requirements' (see Appendix 1) and how they apply to the topic of fields and forces

 Testing your knowledge and understanding

Quick test

Answers to these questions, together with explanations, are in the Answers section which follows Chapter 6.

Select the correct answer to each of the following questions from the four answers supplied. In each case only one of the four answers is correct. Allow about 40 minutes for the 20 questions.

1 The gravitational constant G has base units of

    A $kg\,m^{-2}\,s$ **B** $kg\,m^3\,s^{-2}$ **C** $kg^{-1}\,m^{-2}\,s$ **D** $kg^{-1}\,m^3\,s^{-2}$

2 The Earth may be considered to be a uniform sphere of mass M and radius R. Which of the following equations relates the gravitational constant G and the acceleration of free fall g at the surface of the Earth?

    A $G = R^2/gM$ **B** $G = gR^2/M$ **C** $G = M/gR^2$ **D** $G = gM/R^2$

3 Two similar spheres, each of mass m and radius r, are in contact. One sphere is moved away through a distance $3r$. What is the ratio between the original gravitational attraction between the spheres and the final gravitational attraction between them?

    A $5:2$ **B** $3:1$ **C** $25:4$ **D** $9:1$

4 The gravitational field strength at the surface of a uniform spherical planet is $16\ \mathrm{N\ kg^{-1}}$. The gravitational field strength at the surface of a similar planet having the same density but half the radius would be

    A $4\ \mathrm{N\ kg^{-1}}$ **B** $8\ \mathrm{N\ kg^{-1}}$ **C** $16\ \mathrm{N\ kg^{-1}}$ **D** $32\ \mathrm{N\ kg^{-1}}$

5 Two satellites, of equal mass, are in circular orbits around the Earth. Satellite A has a longer period than satellite B. Which of the following correctly compares the kinetic and potential energies of satellite A with those of satellite B?

    A Both greater
    B Greater kinetic and smaller potential
    C Smaller kinetic and greater potential
    D Both smaller

6 Which of the following statements about a geostationary satellite is not correct?

    A It has an equatorial orbit
    B It has a period of 24 hours
    C It is used for communications
    D The gravitational force on it is negligible

7 Two point charges of magnitudes q_1 and q_2 are a distance d apart. Which of the following changes would double the force acting between them?

    A Doubling either q_1 or q_2
    B Doubling both q_1 and q_2
    C Reducing d to $\frac{1}{2}d$
    D Reducing d to $\frac{1}{4}d$

8 The electrostatic repulsion between two identical charged particles in a vacuum is just balanced by their gravitational attraction. The ratio of charge to mass of each particle is

    A $4\pi\varepsilon_0 G$ **B** $4\pi\varepsilon_0/G$ **C** $\sqrt{(4\pi\varepsilon_0 G)}$ **D** $\sqrt{(4\pi\varepsilon_0/G)}$

9 Which of the following statements about an electric field is not correct?

    A Energy is transferred when a charge moves along an equipotential
    B Field lines show the shape and direction of an electric field
    C Energy is transferred when a charge moves along a field line
    D Field lines and equipotentials cross each other at right angles

10 When a charge of $30\ \mu\mathrm{C}$ is moved between two points, M and N, in a uniform electric field, $150\ \mu\mathrm{J}$ of work is done. The potential difference between M and N is

    A $0\ \mathrm{V}$ **B** $0.2\ \mathrm{V}$ **C** $5\ \mathrm{V}$ **D** $4500\ \mathrm{V}$

11 The electrons in a cathode ray oscilloscope are accelerated from cathode to anode by a potential difference of 2000 V. If this potential difference is increased to 8000 V, the electrons will arrive at the screen with

    A twice the kinetic energy and twice the velocity
    B twice the kinetic energy and four times the velocity
    C four times the kinetic energy and twice the velocity
    D four times the kinetic energy and four times the velocity

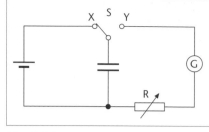

Fig 5.13

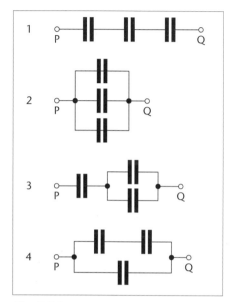

Fig 5.14

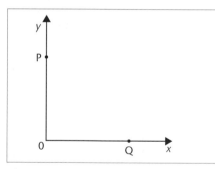

Fig 5.15

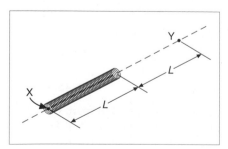

Fig 5.16

12 By moving switch S in the circuit in Figure 5.13 from X to Y, the capacitor is discharged through the variable resistor R and a sensitive meter G.

In one such discharge, the meter showed a maximum deflection of 5 units and took 8 s for the reading to fall to 1 unit. With R set at a decreased value, which of the following is most likely when the switch is moved from X to Y?

A Maximum deflection of 7 units and time taken to fall of 10 s
B Maximum deflection of 7 units and time taken to fall of 5 s
C Maximum deflection of 3 units and time taken to fall of 10 s
D Maximum deflection of 3 units and time taken to fall of 5 s

13 Figure 5.14 shows four possible arrangements of three 1 μF capacitors. Which of the following lists the arrangements in order of increasing capacitance (smallest capacitance first) as measured between P and Q?

A 2 – 1 – 4 – 3   **B** 1 – 3 – 2 – 4   **C** 4 – 1 – 3 – 2   **D** 1 – 3 – 4 – 2

14 The energy stored in a 2 μF capacitor charged to a potential difference of 100 V is

A 0.01 J   **B** 0.02 J   **C** 10 J   **D** 20 J

15 Figure 5.15 shows a horizontal plane O*xy* that contains the perpendicular axes O*x* and O*y*.

Which of the following directions for a current in a straight wire will produce a magnetic field at O in the direction O*x* (i.e. from O towards *x*)?

A Vertically downwards at P
B Vertically upwards at P
C Vertically downwards at Q
D Vertically upwards at Q

16 The magnitude of a uniform magnetic flux density perpendicular to a wire carrying a current can be defined as

A force on wire × length/current
B current × length/force on wire
C force on wire/(current × length)
D current/(force on wire × length)

17 Which of the following statements about the magnetic flux density at the centre of a long current-carrying solenoid is not correct? The magnetic flux density is

A directly proportional to the total number of turns on the solenoid
B inversely proportional to the length of the solenoid
C directly proportional to the current in the solenoid
D inversely proportional to the cross-sectional area of the solenoid

18 The magnetic flux density due to a steady current in the long air-cored solenoid shown in Figure 5.16 is 800 μT at X and 100 μT at Y.

If an identical solenoid is now fixed on the end of the first so as to make it extend as far as Y, and the same current as before passed through the complete solenoid, the magnetic flux density at Y will now be

A 800 μT   **B** 900 μT   **C** 1600 μT   **D** 1800 μT

19 Figure 5.17 shows two aluminium pendula, X and Y, that are identical except for the slots in Y.

Permanent magnets provide strong magnetic fields perpendicular to the regions QRST. Each pendulum is set swinging with the same initial amplitude in the plane of the diagram. Pendulum X comes to rest sooner than pendulum Y because

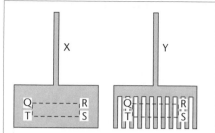

Fig 5.17

A a smaller magnetic field is induced in Y than in X
B Y cuts the magnetic flux more rapidly than does X
C a smaller e.m.f. is induced in Y than in X
D smaller induced currents flow in Y than in X

20 An ideal transformer produces an output voltage of 69 V when connected to the 230 V mains. The secondary coil has 300 turns. The number of primary turns is

A 50   **B** 90   **C** 300   **D** 1000

Worked examples

Study the following worked examples on fields and forces carefully. Make sure you fully understand their answers before attempting the practice assessment questions.

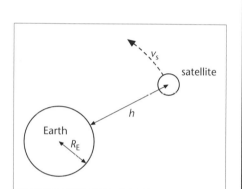

Fig 5.18

Worked example 1

Figure 5.18 (not to scale) shows a satellite of mass m_s in circular orbit at speed v_s around the Earth, mass M_E. The satellite is at a height h above the Earth's surface and the radius of the Earth is R_E.
Using the symbols above, write down an expression for the centripetal force needed to maintain the satellite in this orbit. [2]
Write down an expression for the gravitational field strength in the region of the satellite. [2]
State an appropriate unit for this quantity. [1]
Use your two expressions to show that the greater the height of the satellite above the Earth, the smaller will be its orbital speed. [3]
Explain why, if a satellite slows down in its orbit, it nevertheless gradually spirals in towards the Earth's surface. [2]
(Total 10 marks)
(Edexcel Module Test PH4, June 1997, Q. 2)

Answer:
Centripetal force = mass × centripetal acceleration = mv^2/r ✓
so using the symbols given
centripetal force = $m_s v_s^2/(R_E + h)$ ✓

Gravitational field strength $g = GM/r^2$ ✓
so using the symbols given
$g = GM_E/(R_E + h)^2$ ✓

Unit of gravitational field strength is N kg^{-1} (although m s^{-2} was also allowed) ✓

Gravitational force = mass × gravitational field strength = $m_s \times GM_E/(R_E + h)^2$
Gravitational force provides the required centripetal force
so $m_s \times GM_E/(R_E + h)^2 = m_s v_s^2/(R_E + h)$ ✓
$v_s^2 = GM_E/(R_E + h)$ ✓
so the greater h, the smaller v^2 and the smaller v ✓

If it slows then required centripetal force is less ✓
so that gravitational force > required centripetal force ✓

Unit 5

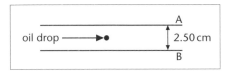

Fig 5.19

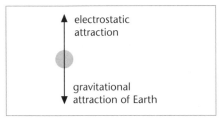

Fig 5.20

Worked example 2

Figure 5.19 shows a positively charged oil drop held at rest between two parallel conducting plates A and B.
The oil drop has a mass of 9.79×10^{-15} kg. The potential difference between the plates is 5000 V and plate B is at a potential of 0 V. Is plate A positive or negative?**[1]**
Draw a labelled free-body force diagram that shows the forces acting on the stationary oil drop. (You may ignore upthrust.) **[2]**
Calculate the electric field strength between the plates. **[2]**
Calculate the magnitude of the charge Q on the oil drop. **[3]**
How many electrons would have to be removed from a neutral oil drop for it to acquire this charge? **[1]**
(Total 9 marks)
(Edexcel Module Test PH4, June 1998, Q. 1)

Answer:
The top plate must be attracting the positively charged drop so it is negative ✓

Free-body diagram (Figure 5.20) showing:
    upwards electrostatic attraction or qE ✓
    downwards gravitational attraction of Earth or mg ✓

Electric field between parallel plates is uniform
for uniform electric fields
$E = V/d = 5000$ V$/(2.50 \times 10^{-2}$ m) ✓
    $= 2.0 \times 10^{5}$ V m^{-1} or N C^{-1} ✓

Since at rest, force up $= qE =$ force down $= mg$ ✓
$q = mg/E = 9.79 \times 10^{-15}$ kg $\times 9.81$ N kg$^{-1}/(2.0 \times 10^{5}$ N C^{-1}) ✓
    $= 4.8 \times 10^{-19}$ C ✓

Electronic charge $= (-) 1.6 \times 10^{-19}$ C
Number of electrons removed $= 4.8 \times 10^{-19}$ C$/(1.6 \times 10^{-19}$ C$) = 3$ ✓

Worked example 3

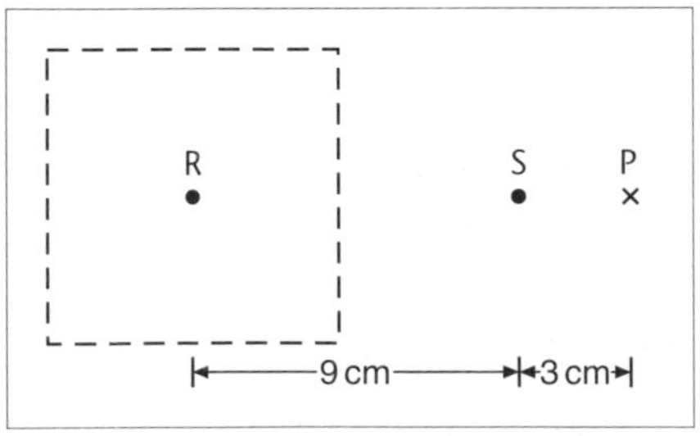

Two long parallel wires R and S carry steady currents I_1 and I_2 respectively in the same direction. Figure 5.21 is a plan view of this arrangement. The directions of the currents are out of the page.
In the region enclosed by the dotted lines, draw the magnetic field pattern due to the current in the wire R alone. **[2]**
The current I_1 is 4 A and I_2 is 2 A. Mark on the diagram a neutral point N where the magnetic flux density due to the currents in the wires is zero. **[2]**
Show on the diagram the direction of the magnetic field at the point P. **[1]**
Calculate the magnitude of the magnetic flux density at P due to the currents in the wires. **[3]**
(Total 8 marks)
(Edexcel Module Test PH4, June 1998, Q. 2)

Answer:
Magnetic field is around the current
so need to add concentric circles (at least 2) around wire R ✓
with arrows in an anticlockwise direction (since current is out of page) ✓

Neutral point occurs where the two magnetic fields are in opposite directions
the anticlockwise fields around R and S are opposite between R and S (B_R up and B_S down) ✓

Fig 5.21

since R has the largest current and the strongest field, N will be closer to S (actually 3 cm from S) ✓

Both anticlockwise magnetic fields act upwards at P
so resultant is upwards – add straight arrow upwards from P (not curved as at a point) ✓

For a long current-carrying wire $B = \mu_0 I / 2\pi r$
at P $B_R = \mu_0 \times 4$ A/$(2\pi \times 12 \times 10^{-2}$ m) and $B_S = \mu_0 \times 2$ A/$(2\pi \times 3 \times 10^{-2}$ m) ✓
$$B_P = B_R + B_S = 6.7 \times 10^{-6}\,\text{T} + 1.3 \times 10^{-6}\,\text{T}\ \checkmark$$
$$= 2.0 \times 10^{-5}\,\text{T}\ \checkmark$$

Worked example 4

Fig 5.22

A large solenoid is 45 cm long and has 72 turns. Calculate the magnetic flux density inside the solenoid when a current of 2.5 A flows in it. **[3]**
A small solenoid is placed at the centre of the large solenoid as shown in Figure 5.22. The small solenoid is connected to a digital voltmeter.
State what would be observed on the voltmeter when each of the following operations is carried out consecutively.
(a) A battery is connected across the large solenoid. **[2]**
(b) The battery is disconnected. **[1]**
(c) A very low frequency alternating supply is connected across the large solenoid. **[1]**
(Total 7 marks)
(*Edexcel Module Test PH4, January 2000, Q. 2*)

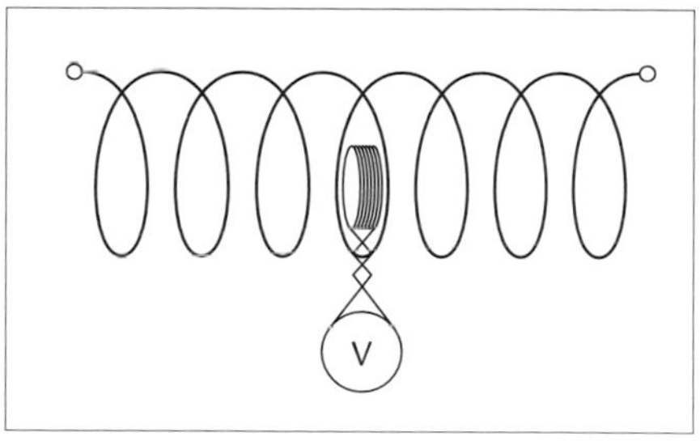

Answer:
For a current-carrying solenoid $B = \mu_0 n I$
n = number of turns per unit length = $N/L = 72/(0.45$ m$) = 160$ m^{-1} ✓
$B = \mu_0 n I = 4\pi \times 10^{-7}\,\text{N A}^{-2} \times 160\,\text{m}^{-1} \times 2.5\,\text{A}$ ✓
$$= 5.0 \times 10^{-4}\,\text{T}\ \checkmark$$

(a) Voltmeter gives a brief reading ✓
(as e.m.f. induced across small solenoid as magnetic flux through it builds up)
and then returns to reading zero ✓
(magnetic flux reaches its maximum value and is no longer changing)

(b) Voltmeter gives a brief reading in the opposite direction and then returns to reading zero ✓
(as opposite e.m.f. induced as magnetic flux falls to zero)

(c) Reading alternates first one way and then the other ✓
(magnetic flux through small solenoid continuously changes direction)

Answers to these questions, together with explanations, are in the Answers section which follows Chapter 6.

Practice questions

The following are typical assessment questions on fields and forces. Attempt these questions under similar conditions to those in which you will sit your actual test.

1 Write a word equation that states Newton's law of gravitation. **[2]**
Mars may be assumed to be a spherical planet with the following properties:

Mass m_M of Mars = 6.42×10^{23} kg
Radius r_M of Mars = 3.40×10^6 m

Calculate the force exerted on a body of mass 1.00 kg on the surface of Mars. **[2]**

For any planet the relationship between g (the free fall acceleration at the surface), the planet's density ρ and its radius R is

$$g = 4\pi\rho GR/3$$

Has Mars a larger, smaller or similar radius to the Earth? Explain your reasoning. **[3]**

(Total 7 marks)

(Edexcel Module Test PH4, June 2001, Q. 3)

2 The gravitational field close to the Earth's surface is often described as uniform. Explain, with the aid of a diagram, what is meant by a uniform gravitational field. **[3]**

Figure 5.23 shows a space rocket R which relies on the kinetic energy it gains in the first few minutes of its flight to carry it from the Earth to the Moon. As the rocket moves further away from the Earth and into the region BC, the decrease in its kinetic energy per kilometre of path gets less. Explain this. **[2]**

The resultant force acting on the rocket is zero at D, 346×10^3 km from the centre of the Earth and 38×10^3 km from the centre of the Moon. Find the ratio of the Earth's mass to the Moon's mass. **[3]**

(Total 8 marks)

(Edexcel Paper 2, June 1993, Q. 6 (part))

3 Calculate the magnitude of the electric field strength at the surface of a nucleus of $^{238}_{92}\text{U}$. Assume that the radius of this nucleus is 7.4×10^{-15} m. **[3]**

State the direction of this electric field. **[1]**

State one similarity and one difference between the electric field and the gravitational field produced by the nucleus. **[2]**

(Total 6 marks)

(Edexcel Module Test PH4, January 2001, Q. 2)

4 A beam of electrons are accelerated from rest through 12 cm in a uniform electric field of strength 7.5×10^5 N C^{-1}. Calculate the potential difference through which the electrons are accelerated. **[2]**

Calculate the maximum kinetic energy in joules of one of these electrons. **[2]**

Calculate the maximum speed of one of these electrons. **[2]**

Draw a diagram to represent the electric field close to an isolated electron. **[2]**

(Total 8 marks)

(Edexcel Module Test PH4, June 2001, Q. 2)

5 State the relationship between current and charge. **[1]**

Two students are studying the charging of a capacitor using the circuit shown in Figure 5.24. The voltmeter has a very high resistance.

The capacitor is initially uncharged. At time zero, one student closes switch S. She watches the milliammeter and continually adjusts the rheostat R so that there is a constant current in the circuit. Her partner records the potential difference across the capacitor at regular time intervals. The graph in Figure 5.25 shows how this potential difference changes with time.

Explain why the graph is a straight line. **[2]**

The capacitor used was 4700 μF. Use the graph to determine the charging current. **[3]**

In order to keep the current constant, did the student have to increase or decrease the resistance of the rheostat as time passed? Explain your answer. **[3]**

Fig 5.23

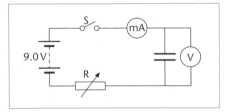

Fig 5.24

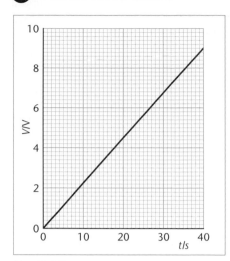

Fig 5.25

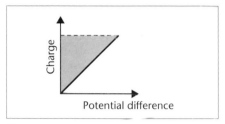

Fig 5.26

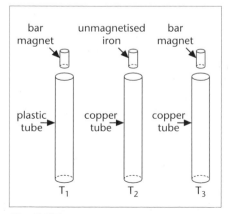

Fig 5.27

The students repeat the experiment, with the capacitor initially uncharged. The initial current is the same as before, but this time the first student forgets to adjust the rheostat and leaves it at a fixed value. Add a second graph to Figure 5.25 to show qualitatively how the potential difference across the capacitor will now change with time. **[2]**

(Total 11 marks)

(Edexcel Module Test PH1, June 2001, Q. 6)

6 Figure 5.26 shows a graph of charge against potential difference for a capacitor.

What quantity is represented by the slope of the graph? **[1]**

What quantity is represented by the shaded area? **[1]**

An electronic camera flash gun contains a capacitor of 100 μF which is charged to a potential difference of 250 V. Show that the energy stored is 3.1 J. **[2]**

The capacitor is charged by an electronic circuit that is powered by a 1.5 V cell. The current drawn from the cell is a steady 0.20 A. Calculate the power from the cell and, from this, the minimum time for the cell to recharge the capacitor. **[3]**

(Total 7 marks)

(Edexcel Module Test PH1, June 2000, Q. 8)

7 The magnitude of the force on a current-carrying conductor in a magnetic field is directly proportional to the magnitude of the current in the conductor. With the aid of a diagram, describe how you could demonstrate this in a school laboratory. **[5]**

At a certain point on the Earth's surface the horizontal component of the Earth's magnetic field is 1.8×10^{-5} T. A straight piece of conducting wire 2.0 m long, of mass 1.5 g, lies on a horizontal wooden bench in an East–West direction. When a very large current flows momentarily in the wire it is just sufficient to cause the wire to lift up off the surface of the bench. Calculate the value of the current and state its direction. **[4]**

What other noticeable effect will this current produce? **[1]**

(Total 10 marks)

(Edexcel Module Test PH4, June 1997, Q. 4)

8 A metal-framed window is 1.3 m high and 0.7 m wide. When closed, it faces due south and opens by pivoting about a vertical edge. The Earth's magnetic field has a horizontal component of 20 μT and a vertical component of 50 μT. Calculate the magnetic flux through the closed window. **[3]**

The window is opened through an angle of 90° in a time of 0.80 s. Calculate the average e.m.f. induced. **[2]**

(Total 5 marks)

(Edexcel Module Test PH4, June 2001, Q. 6)

9 State Lenz's law of electromagnetic induction. **[2]**

An exhibit at a science centre consists of three apparently identical vertical tubes, T_1, T_2 and T_3, each about 2 m long. With the tubes are three apparently identical small cylinders, one to each tube (Figure 5.27).

When the cylinders are dropped down the tubes those in T_1 and T_2 reach the bottom in less than 1 s, whereas that in T_3 takes a few seconds. Explain why the cylinder in T_3 takes longer to reach the bottom of the tube than the cylinder in T_1. **[5]**

Explain why the cylinder in T_2 takes the same time to reach the bottom as the cylinder in T_1. **[2]**

(Total 9 marks)

(Edexcel Module Test PH4, January 1999, Q. 5)

Part ❷   The A2 practical test

 Introduction

This test builds on the practical laboratory skills examined in the Unit PHY3 Practical Test and expects you to demonstrate more advanced skills than in AS. It is based on material from any part of the basic specification. The test consists of three questions each one lasting 25 minutes, although the apparatus may only be used for the first 20 of these. There is a further 15 minutes writing up time at the end giving a total test length of 90 minutes. Each question is worth 16 marks. Question 1 usually contains two exercises of approximately equal length that focus on setting up and using apparatus and recording observations. Question 2 concentrates on using apparatus and evaluating results, with some planning being required. Question 3 concentrates on planning with aspects of analysis and evaluation. At least one question involves drawing a graph, which may involve the use of logarithms. None of the questions will require the use of datalogging apparatus although you may be asked to explain how to set up and use such a device. All advice given in Chapter 3 concerning the AS practical test also applies to that for A2. The following section adds to this advice and gives a sample question of each type. Where possible, you should also practise these questions using the apparatus, which is listed in the questions.

Advice on tackling the A2 practical test

- read carefully through all the advice on tackling the AS practical test given in Chapter 3
- make sure that you know how to use all standard apparatus met with while studying Units 4 and 5 such as signal generators and oscilloscopes
- practise using your calculator to find the logarithm, both natural and base 10, of a number
- practise using your calculator to find the number from a logarithm, both natural and base 10
- remember that taking logarithms of $y = ax^n$ gives $\log y = n \log x + \log a$ which is similar to $y = mx + c$ so the power n is the gradient of the straight line obtained when $\log y$ is plotted against $\log x$
- also, if one quantity decreases exponentially with another, $y = ae^{-kx}$, plotting $\ln y$ against x gives a straight line with a gradient of $-k$, since $\ln y = -kx + \ln a$ which is again similar to $y = mx + c$

Sample A2 practical questions

1 (a) *You are to determine the time taken for the charging current in an RC circuit to halve.*

> Apparatus needed: 4.5 V d.c. supply, 470 μF capacitor with positive terminal clearly marked, 47 kΩ resistor in holder, 100 μA ammeter, five connecting leads

Draw a diagram of the circuit you would use to monitor the current during the charging of a capacitor in series with a resistor from a d.c. power supply. Explain why it is necessary to connect the capacitor into the circuit with the correct polarity. How would you discharge the capacitor while it is still connected to the circuit? **[4]**

> correct series arrangement showing all four components ✓
> with correct polarity of capacitor marked ✓
> capacitor is an electrolytic or to maintain dielectric/insulator ✓
> short out capacitor by connecting spare lead across it ✓

Set up the circuit and determine the time taken for the current to halve. Show all your results below and explain the precautions you took to make your readings as accurate as possible. You are *not* expected to draw a graph. **[4]**

> correct time measured ✓
> and repeated ✓
> for different 'starting' currents ✓
> value between 10 and 20 s (*RC* ln 2 = 15 s) with unit to 2/3 s.f. ✓

(b) *This experiment is concerned with the use of a diffraction grating to observe spectra.*

> Apparatus needed: 12 V, 24 W lamp connected to supply and clamped 75 cm above bench with filament horizontal, 300 line mm⁻¹ diffraction grating clamped a few centimetres below lamp with rulings parallel to filament, A4 sheet of white paper on bench with shorter sides parallel to filament, 10 cm focal length converging lens, tripod supporting a Petri dish containing a very dilute potassium permanganate solution, metre rule, subdued lighting

Place the converging lens on top of the diffraction grating and adjust the height of the grating so that a sharp image of the filament of the lamp is focused on the paper on the bench. Record the distance D from the diffraction grating to the bench. Mark on the piece of paper the limits of the first order visible spectrum on either side of the image of the filament. Sketch the spectra and mark the distance x_R between the two red limits and the distance x_V between the two violet limits. **[3]**

> D, x_R and x_V to nearest mm ✓
> neatly drawn diagram showing filament and spectra ✓
> spectra labelled correctly e.g. violet nearest filament ✓

Calculate the angle θ_R between the first-order red limit of the spectrum and the image of the filament using $\tan \theta_R = x_R/2D$ and hence find $\sin \theta_R$. **[2]**

> correct calculation of $\tan \theta_R$ ✓
> $\sin \theta_R$ found and in range 0.18 to 0.21 ✓

Carefully insert the tripod and dish containing the purple solution so that the light to one of the first order spectra passes through the solution. Describe any changes in the appearance of the spectrum. Explain briefly why the spectrum changes. **[3]**

> black band described or drawn between red and violet ✓
> detail – broad band with narrow red and violet bands ✓
> photons having certain energies are absorbed by solution ✓

(Edexcel Unit Test PHY5, Specimen, Q. A)

Unit 5

Apparatus needed: wooden metre rule labelled A, two 100 g slotted masses, rubber band to secure masses to rule, G-clamp, small block of wood, second metre rule, vernier callipers, digital stopwatch, stand, clamp and boss

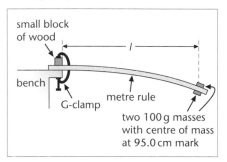

Fig 5.28

2 *You are to investigate how the periodic time of a vibrating cantilever is dependent on the length of the cantilever.*

(a) Set up the apparatus as shown in Figure 5.28 using the metre rule labelled A. The length l should be set to 0.900 m.
Explain carefully how you ensured that the length l was set to 0.900 m. You may add to the diagram if you wish. **[2]**

    use of horizontal second metre rule ✓
    good description of how l was set (words or on diagram) ✓

(b) Displace the end of the rule vertically and determine the period T of the vertical oscillations. **[3]**

    T with unit and averaged from repeated readings ✓
    at least 20 oscillations measured in total ✓
    T within ±0.02 s of supervisor's value ✓

(c) Measure the width b and the thickness d of the clamped rule. Estimate the percentage uncertainty in each of these quantities. **[5]**

    b with unit and within ±0.05 cm of supervisor's value ✓
    d with unit and within ±0.05 cm of supervisor's value ✓
    both b and d repeated ✓
    correct uncertainties (either half-range or ±0.01 cm) ✓
    correct calculations of percentage uncertainties ✓

(d) The Young modulus E of the rule is given by $E = 16\pi^2 M l^3 / b d^3 T^2$ where M = total mass attached to rule. Use your results to calculate a value for E. **[2]**

    substitution of SI values into expression for E ✓
    correct calculation of E with unit ✓

(e) Explain why d would contribute far more than b to the uncertainty in your value for E. **[2]**

    percentage uncertainty in d is larger than that in b ✓
    and is multiplied by 3 (since d^3 in expression) ✓

(f) What additional apparatus would you use to improve the precision in your measurement of d? Estimate the factor by which this would reduce your percentage uncertainty in d. **[2]**

    use a micrometer to measure d ✓
    percentage uncertainty in d reduced by a factor of 10 ✓

    (Edexcel Unit Test PHY5, Specimen, Q. B)

Apparatus needed: 250 ml beaker containing dry sand to within 1 cm of top, table tennis ball

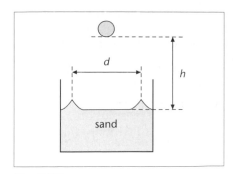

Fig 5.29

3 *You are to plan an investigation of how the diameter of a crater formed in soft sand by a polystyrene sphere is dependent on the impact velocity of the sphere. You are then to analyse a set of data from such an experiment. You may use the sphere and sand provided to observe the crater formation, but you are not required to take any measurements. In addition to the apparatus provided, you may assume that a metre rule, a pair of dividers, a set square and a stand and clamp would be available.*

(a) (i) Which quantity would you vary in order to vary the impact velocity of the sphere? Draw a diagram of your experimental arrangement. Indicate on the diagram any measurements to be taken. Explain how the impact velocity could be found from your measurements. **[4]**

    vary height above sand from which sphere is released ✓
    diagram (Figure 5.29) showing height of bottom of sphere above sand and d ✓
    kinetic energy gained = potential lost ✓
    $\frac{1}{2}mv^2 = mgh$
    $v = \sqrt{(2gh)}$ ✓

(ii) State an assumption which you have to make to determine the impact velocity. How might this assumption affect the range of velocities which you use? **[2]**

    assume that air resistance is negligible ✓
    need to use small heights/velocities to minimise air resistance so limits range used ✓

(iii) The diameter d of the crater is expected to be related to the impact velocity v by an equation of the form $d = kv^n$ where k and n are constants. Describe briefly how you would investigate this relationship experimentally and include an indication of the graph you would plot to investigate the equation. **[3]**

    vary v (by varying h) and measure corresponding d ✓
    plot $\log (d/\mathrm{m})$ against $\log (v/\mathrm{m\ s^{-1}})$ ✓
    taking logarithms of $d = kv^n$ gives $\log d = n \log v + \log k$ should get straight line with a gradient n and intercept $\log (k/\mathrm{m})$ ✓

(b) The following data was obtained in such an investigation.

$v/\mathrm{m\ s^{-1}}$	d/m		
0.44	0.0118		
0.77	0.0168		
1.17	0.0218		
1.47	0.0234		
1.72	0.0275		
2.12	0.0308		

Use the columns provided for your processed data, and then plot a suitable graph to test the equation. **[5]**

$\log (v/\mathrm{m\ s^{-1}})$	$\log (d/\mathrm{m})$
−0.357	−1.93
−0.114	−1.77
0.068	−1.66
0.167	−1.63
0.236	−1.56
0.326	−1.51

plot graph of $\log (v/\mathrm{m\ s^{-1}})$ against $\log (d/\mathrm{m})$ – see Figure 5.30

    correct log values ✓
    suitable scale for graph ✓
    axes labelled with units correctly shown either in table or on graph ✓
    correct plotting ✓
    good best-fit straight line ✓

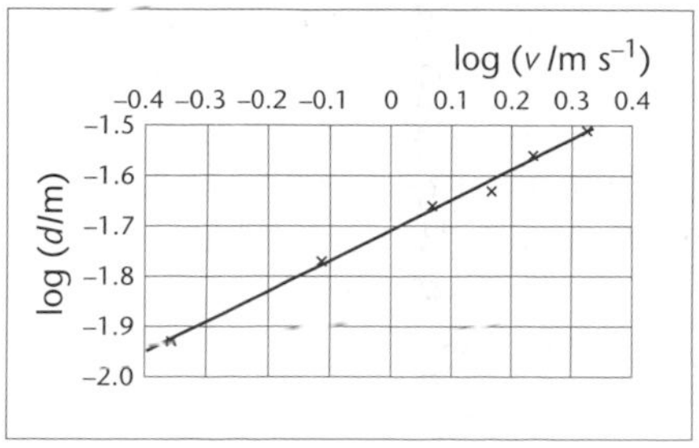

Fig 5.30

(c) Use your graph to determine a value for n. **[2]**

    n = gradient = $-1.50 - (-1.95)/[0.34 - (-0.40)] = 0.45/0.74 = 0.61$
    large triangle used for gradient ✓
    correct calculation giving n to 2 or 3 s.f. and with no unit ✓

Unit 5

6 Synoptic assessment

General Introduction

The synoptic test (PHY6) examines the understanding of physics that you have built up over the full two years of your Advanced GCE course. You will be expected to apply the skills that you acquired from your studies of Units 1, 2, 4 and 5 in answering the questions set. The test consists of four questions; a passage analysis, a question based on the synthesis material in Unit 6 of the specification and two further synoptic questions. The synoptic test is the only one that you answer in a separate answer booklet. Use the mark allocation to help you judge how much detail is expected for each part of your answers. It is also up to you which question you start with. Some candidates spend far too long on the passage analysis to the detriment of their other answers which they then have to rush. They may well have done better if they had started with the other questions. When practising past papers, try working through the questions in different orders to find the order that best suits you.

Part ❶ Passage analysis

Introduction

The passage is usually adapted from a scientific or technological extract from a book or magazine and may not, at first sight, have much to do with the physics that you have studied. However, several of the questions following the passage examine your basic understanding of Advanced GCE physics and about one third of the 32 marks are assigned to these. Other questions might ask you to explain the meaning of terms used in the passage or to perform calculations, either using equations with which you are already familiar or using an equation extracted from the passage.

Things to understand

You need to look back over Units 1, 2, 4 and 5 and check that you understand and can apply the general principles of the physics contained in these units.

Things to learn

There is no new content examined by the passage analysis. However, make sure that you can still remember the equations that are not given in the test for all of the Units.

Advice on tackling the passage analysis

- don't spend ages reading the passage thoroughly
- don't keep reading a particular sentence or paragraph over and over until you understand it as you may not be asked anything about it
- it is best to glance quickly through the passage to get a general feel for its content, highlighting any equations and data that you see
- read all the questions carefully to get your mind thinking about the task ahead
- attempt those questions that have line or paragraph references by reading the appropriate section of the passage
- attempt the remaining questions remembering that not all of these will have answers within the passage
- where you feel your answer is incomplete, leave a space so that you can add to it later if time allows
- keep an eye on the time and don't spend more than the suggested 45 minutes on the passage analysis as there is another 48 marks to be gained in the rest of the paper

Testing your knowledge and understanding

Practice questions

The following are two typical passage analysis questions. Attempt these questions under similar conditions to those in which you will sit your actual test. Try to complete each one in 45 minutes.

Answers to these questions, together with explanations, are in the Answers section which follows Chapter 6.

1 Read the passage and then answer the questions at the end.

An Interesting Moon

In June and September 1996 the Galileo spacecraft made two close approaches to Ganymede, the largest of Jupiter's moons. Measurements made by on-board instruments were beamed back to Earth, while Ganymede's gravitational field was accurately determined by radio-tracking the spacecraft. Two surprising conclusions have now emerged from the data. First, there is a marked decrease in density from the centre of Ganymede to its surface – one of the largest variations of any planetary body in the solar system. Second, Ganymede possesses its own independent magnetic field.

The paths of the Galileo flybys were designed to optimise measurements of Ganymede's gravitational field. The data have allowed the moon's moment of inertia I to be determined, where I is a measure of how the mass of a body is distributed. The results show that

$I/mr^2 = 0.31$

where m and r are Ganymede's mass and radius. This compares with a value of 0.33 for Earth. The gravity measurements also allowed researchers to infer the interior structure of Ganymede. It has an iron-rich core, a silicate mantle and an outer layer of ice, each of comparable thickness. To achieve such a structure, Ganymede must have suffered large-scale melting at some point in its history.

The magnetic field data indicate that Ganymede has a dipole field tilted at about 10° from its spin axis. Ganymede's magnetic field may be internally generated by dynamo action, similar to the process in the Earth. Dynamo theory explains planetary magnetism as a magneto-hydrodynamic phenomenon. In the case of the Earth the magnetic field arises from currents flowing in conducting loops in the molten iron core. The field strength is then limited by the back e.m.f. of the generated field on the motion of the fluid. We must also ask whether there is an energy source to melt the metallic core of Ganymede and to generate motions within it. The core could have been molten when the moon was formed some four billion years ago, and energy released from the decay of long-lived radioactive elements could prevent the liquid metal from freezing. However, it is unlikely that this energy could also maintain vigorous fluid motions.

Another source of energy could be provided by the 'tides' that Jupiter generates in the structure of Ganymede. Such tidal internal heating, particularly at phase boundaries, could initiate or enhance convective motions in the core.

(a)   A bar magnet produces a dipole field. Sketch a diagram showing Ganymede's magnetic field. Label on your sketch the axis about which Ganymede is spinning. **[3]**

(b)   Draw Ganymede as a small circle. Add field lines to show the shape of its external gravitational field in two dimensions.
   Explain how tracking the Galileo spacecraft enables physicists on Earth to determine values for Ganymede's gravitational field strength. **[6]**

(c)   Explain the meaning of the following phrases as used in the passage:
   (i)   radioactive elements (paragraph 3),
   (ii)   back e.m.f. (paragraph 3),
   (iii) convective motions (paragraph 4),
   (iv)   phase boundaries (paragraph 4). **[5]**

(d)   Sketch a graph of the density of the interior of Ganymede against distance from its centre.
   The mean density of the Earth is 5500 kg m^{-3} and its radius is 6400 km. Calculate the moment of inertia I of the Earth. **[6]**

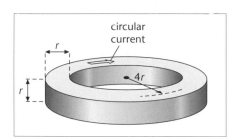

Fig 6.1

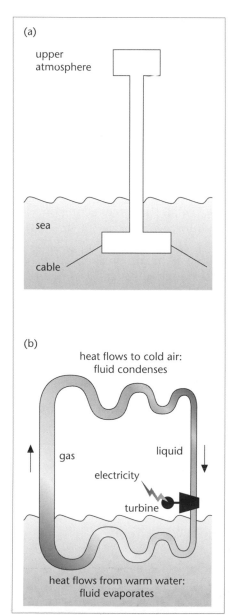

Fig 6.2 (a) A MegaPower tower
(b) The principle of MegaPower

(e) Suppose that the remaining radioactive elements in Ganymede produce a total output power of 1.4×10^{16} W and have a half-life of 1000 million (one billion) years. Calculate the output power of these elements when the moon was first formed. What other energy source is mentioned in the passage? **[4]**

(f) Figure 6.1 shows the current in a conducting loop in Ganymede's molten iron core.

The loop may be considered to be a ring with a square cross-section and dimensions as shown, where $r = 2.0 \times 10^5$ m.
  (i) The resistivity of the material of the core is 1.1×10^{-7} Ω m. Show that the resistance of the ring is 1.4×10^{-11} Ω.
  (ii) What e.m.f. is needed to drive a current of 3.0×10^6 A round the ring?
  (iii) Explain in terms of the laws of electromagnetic induction why it is not possible for this current to fall quickly to zero. **[8]**

(Total 32 marks)
(Edexcel Module Test PH6, January 2000, Q. 1)

2 Read the passage and then answer the questions at the end.

Sky-high tower of power may ride the waves
What is over 7 kilometres tall and dangles into the North Sea? According to researchers in the Netherlands, it could be MegaPower – an enormous power station which they claim may one day be a major source of pollution-free energy. While the scheme may sound crazy, the working cycle that the tower would use to generate electricity is similar to the cycle that underlies hydroelectric power. The turbines in a hydroelectric power station harness the potential energy of water as it falls from a lake or reservoir towards the sea. But this is only half the story. The water gained potential energy when it evaporated from the ocean and rose into the clouds where it cooled down and condensed to form rain that replenished the lakes.

MegaPower would enclose the system inside a giant tower (Figure 6.2a) and replace the water used for hydroelectric power with another fluid. At the top the fluid would condense in the cold of the upper atmosphere. From there it would fall through a turbine to the bottom of the tower where energy from the sea would evaporate the fluid and start the cycle again (Figure 6.2b). It is claimed that the cost of the scheme will be 'within the costs of providing equivalent conventional generating capacity'. One MegaPower installation would have a capacity of 7000 MW. The largest conventional power station in Britain, at Drax in North Yorkshire, is rated at 4000 MW.

In its simplest form, the MegaPower design envisages an enclosing tower 50 metres in diameter with 5 kilometres above the sea. It uses a chemical such as ammonia as the working fluid. The base of the structure would be held in place by three 8-kilometre cables attached to the seabed. Iron or steel cables would break under their own weight, but new materials should be able to cope. To minimise the weight of the tower, it would be built from plastic sandwiched between two skins of aluminium. It would weigh over 400 000 tonnes.

The potential environmental benefits may, one day, offset the development and capital costs and make the project commercially viable. Currently this seems unlikely given the relatively low cost of conventional energy sources and generating methods. Alan Carter, a structural engineer with Amec Process and Energy, was more sceptical: the scheme for MegaPower would not, in his opinion, be economic until hydrocarbon fuels have been exhausted.

(a) Draw a labelled diagram to represent *the cycle that underlies hydroelectric power* (paragraph 1).
Where does hydroelectricity derive its energy? **[7]**

(b) What is the main energy transfer taking place in the turbines of MegaPower as they rotate at a constant rate?
Consider a MegaPower that uses ammonia as the working fluid. Suppose that the liquid ammonia moves through the turbines with a speed of 20 m s^{-1} and that the turbines occupy half the cross-section of the tower (paragraph 3). By considering the energy transfer in 1.0 s, calculate the maximum power output of MegaPower. Take the density of liquid ammonia to be 650 kg m^{-3}. **[7]**

(c) The theoretical efficiency of an ideal heat engine is given by $(T_1 - T_2)/T_1$. In an ideal heat engine, what do the symbols T_1 and T_2 represent?
Assuming that MegaPower acts like an ideal heat engine, what temperatures determine T_1 and T_2 for MegaPower? Suggest a likely value for T_1 and hence estimate T_2 for MegaPower if its calculated maximum efficiency is 0.2. **[6]**

(d) MegaPower, with a mass of over 400 000 tonnes, must be supported by huge upward forces. List all the forces acting on the structure and explain how each force arises. **[3]**

(e) Consider a steel cable of radius r, density ρ and breaking stress σ.
(i) Show that a length of cable h, hanging from a fixed point in the Earth's gravitational field g. will break if $h > \sigma/\rho g$
(ii) Show that the unit of $\sigma/\rho g$ is the metre. **[6]**

(f) What are the environmental benefits of a scheme like MegaPower? Suggest any problems arising from its use other than those discussed in the passage. **[3]**

(Total 32 marks)
(Edexcel Module Test PH6, January 1999, Q. 1)

Synthesis material

Introduction

The content of this section draws together material from different areas of the physics that you studied in the previous Units. The first question in Section II of the synoptic test will examine this material synthesis. As you study the synthesis material, you find that a number of different areas of physics have similarities in the way in which they behave. You compare the stretching of a spring with the charging of a capacitor, radial electric fields with radial gravitational fields, and the discharge of a capacitor with the decay of a radioisotope. You also study how an area of modern physics, the principles of particle accelerators, involves many different areas of physics, such as electric and magnetic fields, circular motion and momentum.

Things to understand

Comparing springs and capacitors

- the force F needed to stretch a spring an extension x is given by $F = kx$ where k is the force constant

- the potential difference V needed to displace a charge Q on a capacitor is given by $V = (1/C) \times Q$ where C is the capacitance

- a graph of F against x is a straight line through the origin of slope k

- a graph of V against Q is a straight line through the origin of slope $1/C$

- a spring with a large force constant requires a larger force for a given extension

- a capacitance with a large capacitance requires a smaller potential difference for a given displacement of charge

- a large capacitance, large Q for given V, is analogous to a weak spring, large x for given F

- energy stored in a stretched spring is the area under a force–extension graph $= \frac{1}{2}Fx$

- energy stored in a charged capacitor is the area under a potential difference–displaced charge graph $= \frac{1}{2}VQ$

Comparing electric and gravitational fields

- electric fields affect all charges

- gravitational fields affect all masses

- electric field strength E is the force per unit charge measured in N C^{-1}

- gravitational field strength g is the force per unit mass measured in N kg^{-1}

- a point/spherical charge produces a radial electric field (Figure 5.3)

- a point/spherical mass produces a radial gravitational field (Figure 5.1)
- the electric field strength of a point/spherical charge obeys an inverse square law, $E = kQ/r^2$
- the gravitational field strength of a point/spherical mass obeys an inverse square law, $g = GM/r^2$
- the constant $k \gg$ the constant G
- electric fields have both attractive and repulsive forces while gravitational fields are always attractive
- an earthed metal container can be used to shield the effects of external electric fields while there is no way of shielding the effects of external gravitational fields

Comparing capacitor discharge and radioactive decay

- the charge displaced on a discharging capacitor decreases exponentially with time, $Q = Q_0 e^{-t/RC}$
- the number of undecayed atoms of a radioactive source decreases exponentially with time, $N = N_0 e^{-\lambda t}$
- discharge current decreases exponentially with time, $I = I_0 e^{-t/RC}$
- activity decreases exponentially with time, $A = A_0 e^{-\lambda t}$
- the discharge current depends on the amount of charge displaced, $I = Q/RC$
- the activity depends on the number of undecayed atoms remaining, $A = \lambda N$
- time constant, RC, is the time taken for the current to decrease to $1/e$ of its original value (Figure 6.3)
- the time taken for the current value to half is the half-life, $\ln 2 \times RC$
- the average time taken for the number of undecayed atoms to halve is the half-life, $t_{\frac{1}{2}} = \ln 2 \times 1/\lambda$
- the time taken for the number of undecayed atoms to fall to $1/e$ of its original value is the time constant, $1/\lambda$
- plotting ln(current) against time gives a straight line of slope $-1/RC$
- plotting ln(number of undecayed atoms) against time gives a straight line of slope $-\lambda$

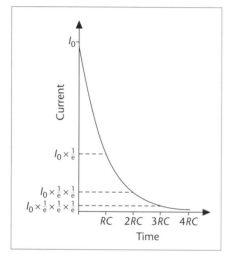

Fig 6.3 *After each time constant RC, the current drops by a factor 1/e*

Mass and energy

- energy has mass, so when a body gains energy it gains mass as well
- increase in mass for an energy gain ΔE is given by $\Delta E/c^2$ where c is the speed of light
- accurate measurements of atomic mass are based on an atom of carbon-12
- the unified mass unit, one twelfth of the mass of a carbon-12 atom, is a convenient unit for calculating mass changes involving sub-atomic particles
- a decrease in mass during a reaction indicates that energy with this amount of mass is released
- splitting up large nuclei into much smaller nuclei releases energy – a process called fission
- a fission reaction is started by the absorption of a neutron and then releases further neutrons, each of which can go on to start another reaction so that a chain reaction can build up (Figure 6.4)

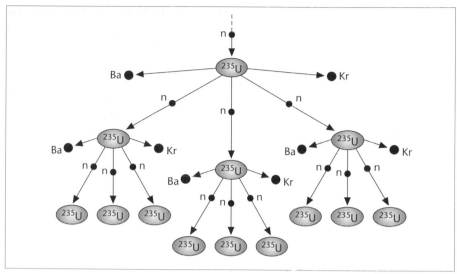

Fig 6.4 *Each fission reaction releases more neutrons*

- joining together of very light nuclei also releases energy – a process called fusion

Linear accelerators

- electric fields are used to accelerate charge particles
- work done qV is transferred to kinetic energy $\frac{1}{2}mv^2$ of accelerated particles
- a Van de Graaff generator is a device that produces a very large voltage used to accelerate charged particles to energies in the MeV range, 10^{-13} J
- particles with higher energies than this are needed to smash atoms and sub-atomic particles
- a linac uses a series of charged plates to continually accelerate charged particles to energies in the GeV range, 10^{-10} J (Figure 6.5)
- the plate separation increases along a linac to allow for the increasing speed of the charged particles
- no matter how much energy a particle gains, its speed can never reach that of light
- the accelerated charged particles are used to bombard other particles in a fixed target

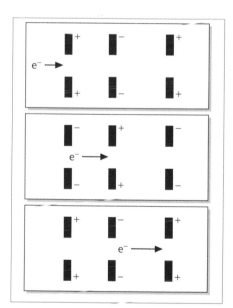

Fig 6.5 *The potential difference across the plates switches repeatedly from positive to negative*

Ring accelerators

- when a charged particle moves at 90° to a magnetic field, a force BQv acts on it perpendicular to both its direction of motion and the magnetic field
- force direction can be found using Fleming's left-hand rule where current direction (second finger) is same as that of positive charges but opposite to that of negative charges
- since force is perpendicular to velocity, no work is done so speed remains the same but direction changes
- charged particles follow a circular path with force BQv providing the centripetal force mv^2/r
- by using magnetic fields, the very long linear path of a linac can be turned into a spiral in a cyclotron or a circle in a synchrotron

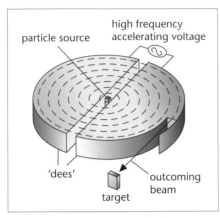

Fig 6.6 *All particles spend the same time in each 'dee'*

- a cyclotron uses an alternating electric field to accelerate charged particles and a magnetic field to make them follow circular paths (Figure 6.6)
- as the charged particles are accelerated every half turn, their path radius in the magnetic field increases
- for speeds up to about 90% of the speed of light, the faster charged particles, despite travelling further, spend the same length of time in each 'dee' as the slower particles
- for such charged particles, the accelerating electric field can have a constant frequency, $f = BQ/2\pi m$
- rather than hitting a fixed target, beams of particles in a synchrotron are sometimes made to collide together from opposite directions
- total momentum of these colliding beams is zero (equal and opposite) so no energy needed for kinetic energy after the collision and thus energy available to create new particles is greater

Detecting particles

- charged particles ionise the atoms of the material through which they pass
- number of ions produced per mm of track depends on charge and speed
- a bubble chamber contains liquid hydrogen: small bubbles form around the trails of ions left by the passage of charged particles
- in a cloud chamber, alcohol droplets from a saturated vapour condense on the trails of ions
- to allow full analysis of the trails (tracks), photographs are taken from a number of angles
- spark and drift chambers contain plates or wires at different high potentials
- currents flow where trails of ions form, and computers use the positions of these currents to reconstruct the paths of the charged particles
- slower particles are involved in more ionising collisions per millimetre of path than fast particles so produce thicker tracks
- alpha particle tracks get thicker towards their ends as the alpha particles slow down
- applying a magnetic field across the paths of the charged particles deflects them along circular paths
- the greater a particle's momentum, the less it is deflected while positive and negative charges are deflected in opposite directions (Figure 6.7)
- as a particle loses kinetic energy through ionising collisions, its momentum and path radius decrease and the particle follows an inward spiral

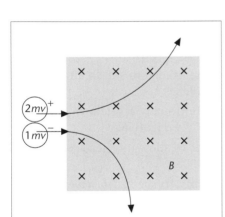

Fig 6.7 *Twice the momentum so twice the path radius*

 ## *Things to learn*

With regards to the amount of learning involved, this part of the specification is somewhat different to the other Units. It is more important that you are familiar with the principles that the synthesis material draws together from different areas of physics. The question set will test your understanding and application of these principles rather than expecting you to recall detailed knowledge about them as your other Unit Tests will already have done this. Therefore you should concentrate on improving your understanding rather than spending time learning detailed knowledge.

 *Checklist*

Before attempting the following questions on the synthesis material, check that you:

- ❏ have compared the stretching of a spring with the charging of a capacitor
- ❏ appreciate the mathematical similarities of the two processes
- ❏ understand that the force constant k of a spring can be compared to the inverse of capacitance $1/C$
- ❏ know that the product of force and displacement and also that of potential difference and charge have units of energy
- ❏ have compared electric and gravitational fields
- ❏ appreciate that all radial fields obey an inverse square law
- ❏ know a number ways in which radial electric and gravitational fields are similar and some of the ways in which they differ
- ❏ have compared capacitor discharge with radioactive decay
- ❏ appreciate the mathematical similarities of the two processes
- ❏ understand why a quantity whose rate of decrease depends on how much of it is there decreases exponentially with time
- ❏ appreciate the constant ratio property of an exponential decay curve and understand how this relates to the concepts of time constant and half-life
- ❏ understand that the decay constant λ can be compared to the inverse of the time constant $1/RC$
- ❏ know how to find the mass difference of a nuclear reaction
- ❏ can calculate the amount of energy released from any mass difference
- ❏ understand how the processes of fission and fusion both release energy
- ❏ appreciate the role played by a uniform electric field in accelerating charged particles
- ❏ know that, for a single pair of electrodes, the energy gained by the charged particles is limited by the maximum accelerating voltage available
- ❏ understand how a linear accelerator uses a series of accelerating voltages to produce beams of very high energy particles
- ❏ appreciate that all particles have the same upper limit to their speed, that of the speed of light
- ❏ know that the main disadvantage of a linac is the extremely long length involved
- ❏ understand how a magnetic field is used to force charged particles into circular paths
- ❏ can use Fleming's left-hand rule to predict the direction in which beams of both positive and negative particles will deflect
- ❏ understand the operating principles of a cyclotron
- ❏ can derive the equation for the frequency of the alternating accelerating voltage, $f = BQ/2\pi m$

□ appreciate that this frequency is constant for all particles with the same charge and mass

□ understand why more energy is available for particle creation when two beams collide head-on than when a single beam collides with a stationary target

□ appreciate that all particle detectors rely on the ionisation produced by charged particles moving through them

□ understand the principles of bubble, cloud, spark and drift chambers

□ can interpret particle tracks in terms of the charge and momentum of the particles producing them

 Testing your knowledge and understanding

Quick test

Answers to these questions, together with explanations, are in the Answers section which follows Chapter 6.

Select the correct answer to each of the following questions from the four answers supplied. In each case only one of the four answers is correct. Allow about 24 minutes for the 12 questions.

1 Which of the following stores the most energy?

    A A spring extended 25 mm elastically by a force of 15 N

    B A capacitor given a 1 mC charge displacement by a potential difference of 9 V

    C A spring of force constant 20 N m^{-1} when extended elastically by 6 cm

    D A 500 µF capacitor charged to a potential difference of 5 V

2 The gravitational and electrostatic interactions between point objects are in some ways analogous. Which of the following correct statements about gravitation is no longer correct if the word 'gravitational' is changed to 'electric' and the word 'mass' is changed to 'positive charge'?

    A The gravitational field of a point mass obeys an inverse square law

    B The magnitude of the gravitational field due to a point mass is proportional to the mass

    C The direction of the gravitational field is towards the mass

    D The gravitational field of a point mass is radial

3 The table lists pairs of particles, a type of force between them and gives their distance apart.

	particles	type of force	distance apart
1	2 protons	electrical	1 m
2	2 protons	gravitational	1 m
3	2 protons	gravitational	0.1 m
4	2 electrons	electrical	0.1 m

Which of the following puts the forces between the particles in order of increasing magnitude?

    A 2134   **B** 1423   **C** 3214   **D** 2314

4 The initial charge on a capacitor is 10 µC. The capacitor discharges through a resistance of 1.0 kΩ. The time constant for the discharge circuit is 1.0 ms. Which of the following statements is correct?

    A The capacitance is 10 µF

    B The capacitor was charged to a potential difference of 10 V

    C The charge remaining on the capacitor after 2.0 ms is 2.5 µC

    D The initial current is 1.0 mA

5 A radioactive material has a half-life of 5×10^7 s. The time constant, in seconds, for its decay is given by

    A $5 \times 10^7 \times \ln 2$

    B $5 \times 10^7 / \ln 2$

    C $\ln 2 / (5 \times 10^7)$

    D $1/(5 \times 10^7)$

6 Given that the atomic mass of $^{14}_{7}\text{N}$ is 14.003074 u and that the sum of the atomic masses of $^{1}_{1}\text{H}$ and $^{13}_{6}\text{C}$ is 14.011179 u, it would be reasonable to suppose that the nuclear reaction

$$^{1}_{1}\text{H} + {}^{13}_{6}\text{C} \rightarrow {}^{14}_{7}\text{N}$$

    A can only happen if there is a net supply of energy

    B could not take place at all

    C must involve the emission of a further uncharged atomic particle

    D will result in the emission of energy

7 In a nuclear physics experiment, a high-energy beam of $^{32}_{16}\text{S}$ (sulphur) bombards a target consisting of $^{94}_{42}\text{Mo}$ (molybdenum) atoms. Many heavier nuclei are produced, each accompanied by the emission of a number of fragments such as protons and neutrons. Which of the following nuclei produced could be associated with the emission of only a single alpha particle?

    A $^{122}_{50}\text{Sn}$ **B** $^{126}_{50}\text{Sn}$ **C** $^{122}_{56}\text{Ba}$ **D** $^{123}_{56}\text{Ba}$

8 An alpha particle moves in a uniform magnetic field of strength 2.0 T at 1.6×10^7 m s^{-1} perpendicular to the field. The electronic charge is 1.6×10^{-19} C. Which of the following gives the force, in newtons, on the alpha particle?

    A 1.0×10^{-11} **B** 5.1×10^{-12} **C** 4.0×10^{-26} **D** 2.0×10^{-26}

9 An electron travelling at a constant speed passes into a uniform magnetic field perpendicular to its direction of motion and moves in a circular path of radius r. If the flux density of the field were halved and the speed of the electron doubled, the new radius would be

    A $\frac{1}{4}r$ **B** r **C** $2r$ **D** $4r$

10 A particle P with a charge Q, mass m and constant speed v crosses the line QR into a region where a uniform magnetic field B causes it to move in a semicircular path of radius r as shown in Figure 6.8. The particle takes a time t to travel in the semicircle from Q to R. How long would a particle, identical with P except that it has speed $2v$, projected similarly from Q take to re-cross the dashed line QR.

    A $\frac{1}{4}t$ **B** $\frac{1}{2}t$ **C** t **D** $2t$

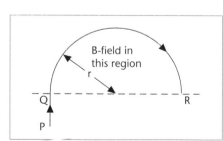

Fig 6.8

11 All particle detectors work because the charge particles moving through them cause

    A ionisation   **B** evaporation   **C** excitation   **D** condensation

12 A nuclear physicist obtains a bubble chamber photograph of the path of a charged particle. The track spirals inwards. Which of the following is the reason for the inward spiral?

    A The charge on the particle is decreasing

    B The speed of the particle is decreasing

    C The magnetic flux density is decreasing

    D The mass of the particle is increasing

Worked examples

Study the following worked examples on the synthesis material carefully. Make sure you fully understand their answers before attempting the practice assessment questions.

Worked example 1

A 2 µF capacitor is charged to a potential difference of 9 V. Calculate the charge stored. **[3]**

The charged capacitor is then connected across a 3 MΩ resistor. Calculate the initial current in the circuit. **[2]**

Calculate the time constant of the circuit. **[2]**

Draw a graph of how the current would vary over the first 24 s. Label the axes and include appropriate scales. **[4]**

(Total 11 marks)

(*Edexcel Module Test PH4, June 2000, Q. 2*)

Answer:
$Q = CV$ ✓
$\quad = 2 \times 10^{-6}\,\text{F} \times 9\,\text{V}$ ✓
$\quad = 1.8 \times 10^{-5}\,\text{C}$ ✓

Initial current depends only on initial potential difference and resistance
$I = V/R = 9\,\text{V}/(3 \times 10^6\,\Omega)$ ✓
$\quad\quad = 3 \times 10^{-6}\,\text{A}$ ✓

Time constant $= CR$ ✓
$\quad\quad = 2 \times 10^{-6}\,\text{F} \times 3 \times 10^6\,\Omega = 6\,\text{s}$ ✓

Graph (Figure 6.9) showing:
$\quad$ concave falling curve ✓
$\quad$ starting from 3 µA at $t = 0$ ✓
$\quad$ current falls to approximately 1 µA after 6 s ✓
$\quad$ graph extends to 24 s and does not cut/meet time axis ✓

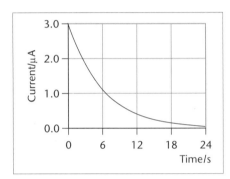

Fig 6.9

Worked example 2

Calculate the energy released by the following reaction:
$$^2_1\text{H} + ^2_1\text{H} \longrightarrow ^3_2\text{He} + ^1_0\text{n}$$
Data:$\quad$ Mass of $^2_1\text{H} = 2.014\,458\,\text{u}$
$\quad\quad\quad$ Mass of $^3_2\text{He} = 3.016\,868\,\text{u}$
$\quad\quad\quad$ Mass of $^1_0\text{n} = 1.009\,036\,\text{u}$ **[4]**

State one similarity and one difference between nuclear fission and nuclear fusion. **[2]**

(Total 6 marks)

(*Edexcel Module Test PH2, January 1997, Q. 3*)

Answer:
Total mass of particles before $= 2 \times 2.014458\,\text{u} = 4.028916\,\text{u}$
Total mass of particles after $= 3.016868\,\text{u} + 1.009036\,\text{u} = 4.025904\,\text{u}$
Mass of energy released $= 4.028916\,\text{u} - 4.025904\,\text{u} = 0.003012\,\text{u}$ ✓
$\quad\quad\quad\quad\quad\quad\quad\quad = 0.003012 \times 1.66 \times 10^{-27}\,\text{kg} = 5.00 \times 10^{-30}\,\text{kg}$ ✓

energy released $= c^2 \Delta m = (3 \times 10^8\,\text{m s}^{-1})^2 \times 5.00 \times 10^{-30}\,\text{kg}$ ✓
$\quad\quad\quad\quad\quad = 4.50 \times 10^{-13}\,\text{J}$ ✓

Answers to these questions, together with explanations, are in the Answers section which follows Chapter 6.

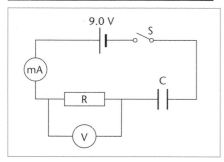

Fig 6.10

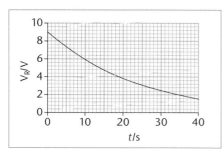

Fig 6.11

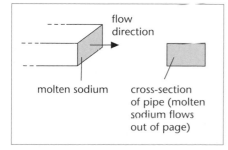

Fig 6.12

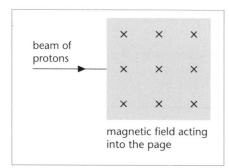

Fig 6.13

Practice questions

The following are typical assessment questions on the synthesis material. Attempt these questions under similar conditions to those in which you will sit your actual test.

1   A circuit is set up as shown in Figure 6.10.
    At $t = 0$ switch S is closed. Readings of the potential difference across the resistor are taken at regular intervals and the graph shown in Figure 6.11 is obtained.
    Use the graph to estimate the time constant for this circuit.   **[2]**
    The initial current $I_0 = 0.19$ mA. Calculate the resistance of resistor R.   **[2]**
    Hence calculate the capacitance of the capacitor C.   **[2]**
    Add to the graph in Figure 6.11 a line showing how the potential difference across the capacitor varies with time over the same period.   **[2]**
    (Total 8 marks)
    (Edexcel Module Test PH4, January 2001, Q. 4)

2   An example of a nuclear fission reaction is given by the equation
    $$^{1}_{0}n + ^{235}_{92}U \rightarrow ^{144}_{56}Ba + ^{92}_{36}Kr$$
    Data: mass of $^{235}_{92}U$ = 235.04394 u
    mass of $^{144}_{56}Ba$ = 143.92285 u
    mass of $^{92}_{36}Kr$ = 91.92627 u
    mass of $^{1}_{0}n$ = 1.00870 u
    Use this data to calculate the energy released in joules for each fission.   **[4]**
    (Total 4 marks)
    (Edexcel Module Test PH2, June 2001, Q. 2)

3   Figure 6.12 shows a beam of protons about to enter a region where a magnetic field acts at right angles to the beam.

    Draw the path of the protons after entering the magnetic field.   **[1]**
    Add to Figure 6.12 the path of a beam of protons travelling at a lower speed which enters the magnetic field from the opposite direction.   **[2]**
    The fact that a moving charge experiences a force in a magnetic field is used in an electromagnetic pump which moves molten sodium through pipes in a nuclear power station. The pipes are situated in a magnetic field and a direct current is passed through the sodium. Figure 6.13 shows a pipe of rectangular cross-section containing molten sodium.

    Show, on the right-hand diagram of the pipe's cross-section, the direction in which a current would have to flow through the sodium together with the orientation of the magnetic field if the molten metal is to flow out of the page.   **[3]**
    (Total 6 marks)
    (Edexcel Module Test PH4, June 2000, Q. 3)

Part ❸ Synoptic questions

Introduction

The last two question of Unit Test PHY6 are the synoptic questions. Fortunately, there is no new content to learn for these. However, each question examines your understanding of the principles from more than one of your previous units so your preparation for these questions will be similar to that for the passage analysis. You need to look back over Units 1, 2, 4 and 5. Check that you understand and can apply the general principles of the physics contained in these units and make sure that you can still remember the equations that are not given in the test for all of the Units.

Advice on tackling the synoptic questions

● if leaving these until last, make sure you leave sufficient time for them

● look at every part of each question – don't let a difficult part put you off easier parts that follow

● leave plenty of space to return to any parts that you miss out and to add extra detail to any partial answers

● keep your answers concise and to the point

● don't write lengthy essays

● where possible use sketches to support your answers, a good labelled sketch can replace several sentences

---

Answers to these questions, together with explanations, are in the Answers section which follows Chapter 6

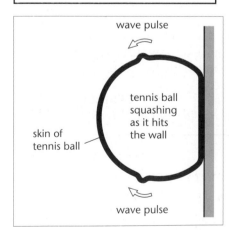

Fig 6.14

Practice questions

The following are typical synoptic questions. Attempt these questions under similar conditions to those in which you will sit your actual test.

1 A tennis ball, moving horizontally at a high speed, strikes a vertical wall and rebounds from it.

  (a) Describe the energy transfers which occur during the impact of the ball with the wall. **[2]**

  (b) During the impact the volume of gas in the tennis ball decreases from V to $0.84V$. The initial pressure of the gas in the ball is 116 kPa.

    (i) Show that the increase in pressure which this produces is 22 kPa assuming that the temperature of the gas remains constant.

    (ii) In practice the gas pressure increases by 32 kPa, since the temperature changes as the ball is squashed. Explain this increase in pressure in molecular terms. **[6]**

  (c) As a result of the impact, a ring-like wave pulse travels along the 'skin' of the tennis ball as shown in Figure 6.14. The speed c of this pulse in the skin of the ball is given by

$$c = \sqrt{(2\pi a^3 \Delta p/m)}$$

where m is the mass of the tennis ball, a its radius and Δp is the difference between the pressure of the gas inside the ball and atmospheric pressure.

(i) The time of contact t_c of the ball with the wall is the time taken for this wave pulse to move to the non-impact side of the ball (i.e. half way round its circumference) and back again. For a ball of mass 57.5 g and radius 33.5 mm, estimate t_c. Take atmospheric pressure to be 100 kPa and refer to the pressure data in part (b). **[5]**

(ii) Suggest a method by which you could measure the time of contact of the tennis ball and the wall to a precision of about a millisecond. **[3]**

(Total 16 marks)

(Edexcel Module Test PH6, June 2001, Q. 3)

2 (a) The shaded square in Figures 6.15, 6.16 and 6.17 represents a piece of resistance paper. The surface of the paper is coated with a conducting material. In Figure 6.15 two metal electrodes E_1 and E_2 are placed on the resistance paper and connected to a battery.

(i) Sketch the electric field in the region between E_1 and E_2.

(ii) E_1 and E_2 are 15 cm apart. What is the strength of the electric field at X, a point halfway between them?

(iii) Add and label three equipotential lines in the region between E_1 and E_2. **[7]**

(b) Figure 6.16 shows two 470 Ω resistors and a milliammeter connected to the initial arrangement. The other side of the milliammeter is connected to a metal probe which makes contact with the surface of the resistance paper.

(i) The metal probe is moved over the resistance paper surface. When the probe is at X the milliammeter registers zero. State the potential at X and explain why the milliammeter registers zero.

(ii) Describe how you would adapt the apparatus to find the potentials at other points on the resistance paper. **[5]**

(c) The resistance of a square piece – a tile – of the resistance paper is given by $R = \rho/t$, where ρ is the resistivity and t the thickness of the material forming the conducting layer.

(i) By considering the square of side x shown in Figure 6.17, prove that $R = \rho/t$, i.e. that the resistance of the tile is independent of the size of the square.

(ii) Calculate the resistivity of a material of thickness 0.14 mm which has a resistance of 1000 Ω for a square of any size. **[4]**

(Total 16 marks)

(Edexcel Module Test PH6, June 2001, Q. 4)

3 Figure 6.18 shows a range of wave speeds for water waves.

(a) Tsunamis, sometimes called tidal waves, are caused by events like underwater earthquakes. Their wavelength λ is related to

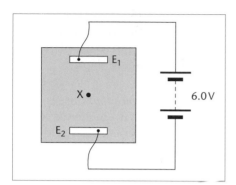

Fig 6.15

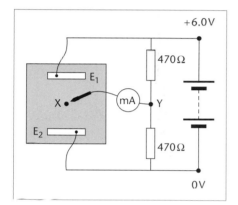

Fig 6.16

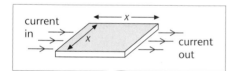

Fig 6.17

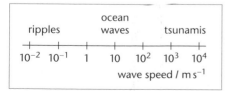

Fig 6.18

the time period of the waves T by the equation

$$\lambda = T\sqrt{(gd)}$$

where d is the depth of water in which they are travelling and g is the value for the gravitational field strength. Show that this expression is homogeneous with respect to units. What is the speed of a wave of time period 18 min in water of depth 2200 m? [5]

(b) Describe an experiment with ripples which you would use to demonstrate the principle of superposition. With reference to your demonstration, explain what measurements you would need to make in order to calculate a value for the wavelength of the ripples. [6]

(c) The average power of the waves arriving at a 200 m long ocean beach is 2.0 MW.

(i) Explain why the beach does not become very hot because of this input power of 2.0 MW.

(ii) Suggest two difficulties which would arise in trying to store this energy, or a fraction of it, for future use.

(iii) The efficiency with which this wave power could be transferred to electricity is 1.6%. Calculate how many kilowatt-hours of electrical energy would be produced in 24 h. [5]

(Total 16 marks)

(Edexcel Module Test PH6, January 1999, Q. 3)

4 (a) In an oscilloscope, N electrons each of charge e hit the screen each second. Each electron is accelerated by a potential difference V.

(i) Write down an expression for the total energy of the electrons reaching the screen each second.

(ii) The power of the electron beam is 2.4 W. When the oscilloscope is first switched on the spot on the glass screen is found to rise in temperature by 85 K during the first 20 s. The specific heat capacity of glass is 730 J kg^{-1} K^{-1}. Calculate the mass of glass heated by the electron beam. State two assumptions you have made in your calculation. [7]

(b) Outline how, in principle, you would measure the specific heat capacity of glass. You may use a lump of glass of any convenient shape in your experiment. What difficulties might lead to errors? [5]

(c) The oscilloscope is now used to investigate the 'saw-toothed' signal from a signal generator. The trace shown in Figure 6.19 is obtained.

The Y-gain control is set at 0.2 V per division and the time-base control at 100 µs per division.

(i) Calculate the frequency of the saw-toothed signal.

(ii) What is the rate of rise of the signal voltage during each cycle? [4]

(Total 16 marks)

(Edexcel Module Test PH6, January 1999, Q. 5)

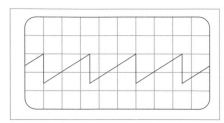

Fig 6.19

Answers

Unit 1 Mechanics and radioactivity

Part 1 Mechanics

Quick test

1 C The six base quantities required for this course are length, mass, time, current, temperature interval and amount of substance and their associated base units are metre, kilogram, second, ampere, kelvin and mole

2 D c measured in $m\,s^{-1}$; ρ measured in $kg\,m^{-3}$
Equation $c = \sqrt{(k/\rho)}$ can be written as $c^2 = k/\rho$
so $\quad k = c^2\rho$
Units of
$k = (m\,s^{-1})^2 \times kg\,m^{-3} = m^2\,s^{-2}\,kg\,m^{-3} = kg\,m^{-1}\,s^{-2}$

3 C Density = mass/volume
Volume = mass/density = $50\,kg/(2500\,kg\,m^{-3}) = 0.02\,m^3$

4 A As body falls, its height decreases with time ... so **A** or **B**
As body accelerates, its velocity increases with time
Slope (= velocity) of graph A increases with time
[slope of graph B represents a constant velocity]

5 B Speed = distance/time
Reading data from the graph gives
speed = $10\,m/(5\,s) = 2\,m\,s^{-1}$

6 C $v = ?$ $\quad u = 10\,m\,s^{-1}$ $\quad x = 120\,m$ $\quad a = 1.6\,m\,s^{-2}$
Select equation containing only these quantities
$v^2 = u^2 + 2ax = (10\,m\,s^{-1})^2 + (2 \times 1.6\,m\,s^{-2} \times 120\,m)$
$v^2 = 100\,m^2\,s^{-2} + 384\,m^2\,s^{-2} = 484\,m^2\,s^{-2}$
$v = \sqrt{(484\,m^2\,s^{-2})} = 22.0\,m\,s^{-1}$

7 C Ball continues to move horizontally at $1\,m\,s^{-1}$ after release
Camera moves horizontally at same speed as falling ball so only records ball's vertical motion
[answer would be **B** if camera was not moving]

8 D Weight ... Earth pulls rocket down with a gravitational force
third law force pair $\quad$ same type (gravitational)
$\qquad\qquad\qquad\qquad$ exchange the two bodies
$\qquad\qquad\qquad\qquad$ reverse the direction
so ... rocket pulls Earth up with a gravitational force

9 D Box is in equilibrium so force Q = forces P + R, ∴ not **A**
A free-body force diagram shows only the forces acting on that body i.e. on the box
[forces **B** and **C** would appear on a free-body force diagram of this situation for the Earth]

10 A For equilibrium, forces must cancel so resultant force = 0
[forces in **B** are in vertical equilibrium but have a resultant towards the right; forces in both **C** and **D** cannot cancel as there are no downward components to cancel the upward ones]

11 D About any point, anticlockwise moments = clockwise moments
About point P ...$F_2 \times d_1 = F_3 \times (d_1 + d_2)$ $\quad$ hence **D**
About 'point F_2'...$F_1 \times d_1 = F_3 \times d_2$
About point Q ...$F_1 \times (d_1 + d_2) = F_2 \times d_2$

12 C

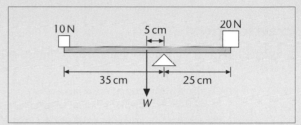

Fig A.1 The weight W of the rod acts at its centre

Weight W acts down from centre of uniform rod (Figure A.1)
Taking moments about the pivot:
$\quad$ clockwise $\qquad\qquad 20\,N \times 25\,cm = 500\,N\,cm$
$\quad$ anticlockwise $\qquad (10\,N \times 35\,cm) + (W \times 5\,cm)$
So $\quad 350\,N\,cm + (W \times 5\,cm) = 500\,N\,cm$
$\quad\quad W \times 5\,cm = 500\,N\,cm - 350\,N\,cm = 150\,N\,cm$
$\quad\quad W = 150\,N\,cm/5\,cm = 30\,N$

13 B As moving at a steady speed, resultant force must be zero
Not **A** as lift $\times \sin\theta$ is horizontal
Not **C** as both thrust and drag also have vertical components

14 C Either $\quad$ acceleration $a = F/m = 5\,N/4\,kg = 1.25\,m\,s^{-2}$
$\qquad v = ?$ $\quad u = 0\,m\,s^{-1}$ $\quad a = 1.25\,m\,s^{-2}$ $\quad t = 8\,s$
$\qquad$ speed $v = u + at = 0 + (1.25\,m\,s^{-2} \times 8\,s)$
$\qquad\qquad\qquad = 10\,m\,s^{-1}$
Or $\qquad$ impulse $Ft = 5\,N \times 8\,s = 40\,N\,s$
$\qquad$ so change in momentum = $40\,kg\,m\,s^{-1}$
$\qquad mv = 40\,kg\,m\,s^{-1}$ (since $mu = 0$)
$\qquad v = 40\,kg\,m\,s^{-1}/m = 40\,kg\,m\,s^{-1}/4\,kg = 10\,m\,s^{-1}$

15 B Impulse = force $\times$ time = area under a force–time graph

16 A Momentum is conserved in every collision where no external force acts

17 B Kinetic energy transfers into gravitational potential energy
First object $\qquad mgh = \frac{1}{2}mv^2$
$\qquad\qquad\qquad h = v^2/2g$
Second object $\quad (\frac{1}{2}m)gh_2 = \frac{1}{2}(\frac{1}{2}m)(\frac{1}{2}v)^2 = mv^2/16$
$\qquad\qquad\qquad h_2 = v^2/8g = \frac{1}{4} \times v^2/2g = \frac{1}{4} \times h$

18 A Either $\quad$ acceleration $a = F/m = -27\,N/0.15\,kg$
$\qquad\qquad\qquad\qquad\qquad = -180\,m\,s^{-2}$
$\qquad v = 0\,m\,s^{-1}$ $\quad u = ?$ $\quad x = 0.4\,m$ $\quad a = -180\,m\,s^{-2}$
$\qquad v^2 = u^2 + 2ax$ so $u^2 = v^2 - 2ax$
$\qquad u^2 = 0^2 - (2 \times -180\,m\,s^{-2} \times 0.4\,m) = 144\,m^2\,s^{-2}$

$u = \sqrt{(144 \ m^2 \ s^{-2})} = 12 \ m \ s^{-1}$

Or   work done Fx = change in kinetic energy $\frac{1}{2}mu^2$

$u^2 = 2Fx/m = 2 \times 27 \ N \times 0.4 \ m/0.15 \ kg$

$= 144 \ m^2 \ s^{-2}$

$u = \sqrt{(144 \ m^2 \ s^{-2})} = 12 \ m \ s^{-1}$

19 D   Elastic collision so momentum and kinetic energy conserved

Initial momentum are $+ mv$ and $- mv$ (opposite directions), so total momentum before and after collision is zero

Initial kinetic energies are $\frac{1}{2}mv^2$ and $\frac{1}{2}mv^2$ (energy is a scalar), so total kinetic energy before and after collision is mv^2

20 D   Power = energy transferred/time taken

Since girl is gaining gravitational potential energy

power = mgh/t = 50 kg $\times$ 10 m s^{-2} $\times$ 5 m/4 s = 625 W

Practice questions

1   Water has maximum density at $\theta = 4.1$ °C (range 4.0 °C to 4.2 °C) ✓

Sketch graph showing: opposite shape ✓

                             mirror image with maximum density at about 4.1 °C ✓

Clarity mark – answer needs to make sense on first reading and attempt to address the question asked ✓

Method of measuring volume e.g. measuring cylinder/syringe/test tube (NOT beaker) ✓

Method of heating e.g. freezing and allowing to warm up in air/water bath from 0 °C to 10 °C (NOT Bunsen) ✓

Magnification of small volume change: e.g. test tube fitted with bung and narrow tube in which change in level is more noticeable ✓

2   Measure exact length of truck ✓

Light gate, attached to a timer, placed above point A (Figure A.2) ✓

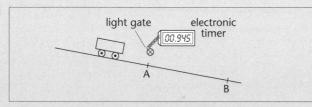

Fig A.2 *Truck obstructs beam of light gate as it passes the point A*

Timer records how long truck takes to pass through light beam ✓

Average speed = length of truck/recorded time ✓

$v = 1.64 \ m \ s^{-1}$ $u = 1.52 \ m \ s^{-1}$ $x = 1.2 \ m$ $a = ?$

$v^2 = u^2 + 2ax$

$a = (v^2 - u^2)/2x$

$= [(1.64 \ m \ s^{-1})^2 - (1.52 \ m \ s^{-1})^2]/(2 \times 1.2 \ m)$ ✓

Examiner's comments

Alternatively, you could first divide distance AB by the truck's *average* speed (1.58 m s^{-1}) between A and B to find how long it takes to go from A to B (0.76 s) and then use $a = (v - u)/t$.

$a = 0.379 \ m^2 \ s^{-2}/2.4 \ m = 0.158 \ m \ s^{-2}$ ✓

3   **(i)**   displacement = area under graph (as car moves along a straight line)

Displacement after 1.5 s = area under graph to the left of t = 1.5 s ✓

$= \frac{1}{2} \times 32 \ m \ s^{-1} \times 1.5 \ s = 24 \ m$ ✓

Allowable range ... 22 m to 26 m

Examiner's comments

Where possible, try to approximate the area to a known shape, a right-angled triangle in this case (Figure A.3)

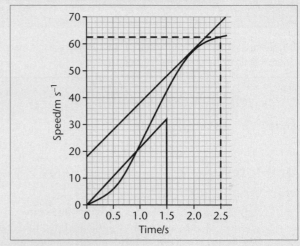

Fig A.3 *As the car's acceleration is varying, you cannot use the equations of motion*

(ii)   Acceleration = gradient of tangent to curve

Tangent drawn at t = 2.0 s (Figure A.3) ✓

Gradient = (70 m s^{-1} – 18 m s^{-1})/2.6 s

$= 52 \ m \ s^{-1}/2.6 \ s = 20 \ m \ s^{-2}$ ✓

Allowable range ... 15 m s^{-2} to 24 m s^{-2}

Examiner's comments

A 'see-through' ruler is essential when judging where to draw both the triangle and the tangent.

(iii)   At t = 2.5 s, v = 62.5 m s^{-1} (allowable range ... 62 m s^{-1} to 63 m s^{-1})

Kinetic energy = $\frac{1}{2}mv^2$

$= \frac{1}{2} \times 420 \ kg \times (62.5 \ m \ s^{-1})^2$ ✓

$= 8.2 \times 10^5 \ J$ ✓

Allowable range ... 8.0×10^5 J to 8.4×10^5 J

4

	Description of force	Body which exerts force	Body the force acts on
Force A	Gravitational	Earth	Child
Force B	Reaction/ Contact ✓	Earth/ **AND** Ground	Child ✓
Force C	Gravitational ✓	Child **AND**	Earth ✓

A and B are equal as the child is at rest/in equilibrium ✓
Action and reaction are equal and opposite/Newton's third law ✓

To jump, child must push down/increase force D ✓
so force B increases ✓
and force B > force A/upward resultant force on child ✓

5   Each person supports half the weight = 200 N ✓

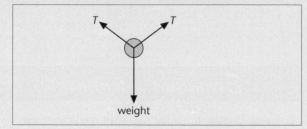

Fig A.4 *Use a 'blob' to represent the container*

Free-body force diagram (Figure A.4) showing:
two tensions ✓
Weight/pull of Earth/gravitational force/mg/400 N ✓
('gravity' on its own would not normally get this mark, so avoid it)

Attempt to resolve vertically ✓
$2T \sin 40° = 400$ N ✓

$T = 310$ N ✓
Applied force is smaller ✓
Forces are only upwards/campers are not pulled sideways ✓

Tension would be greater/larger sideways force ✓

6   Momentum = mass × velocity (NOT mv) ✓

Newton's second law used in line 3 ✓

Newton's third law used in line 2/lines 1 and 2 ✓

Assumption: no external forces/no friction ✓
should appear in line 3 (ONLY if previous mark awarded) ✓

Suitable collision described and specific equipment to measure velocities (e.g. light gates) ✓
Measure velocities before and after collision ✓
How velocities are calculated (e.g. measured distance/measured time) ✓
Measure masses/use known masses/use equal masses ✓
Calculate total initial and final momentum and compare ✓   **Max 4**

7   Momentum = mv = 50 kg × 30 m s^{-1} ✓
                = 1500 kg m s^{-1} or 1500 N s ✓

Either
$Ft = \Delta mv$ ✓
$F \times 0.070$ s = 1500 N s ✓
$F = 21$ kN ✓
or
$F = ma$ ✓
$F = 50$ kg × 30 m s^{-1}/(0.070 s) ✓
$F = 21$ kN ✓
Several forces could contribute to the resultant force on the driver e.g. force from airbag/seat/steering wheel ✓

8   Time = distance/speed = $2\pi r/v = 2\pi \times 60$ m/(0.20 m s^{-1}) ✓
            = 1900 s (1884 s or 31.4 min) ✓

Velocity changes from 0.20 m s^{-1} upwards to 0.20 m s^{-1} downwards ✓
so velocity change = 0.40 m s^{-1} (downwards) ✓

GPE = mgh = 80 kJ ✓
$m = 80 \times 10^3$ J/(9.81 m s^{-2} × 120 m) ✓
$m = 68$ kg ✓

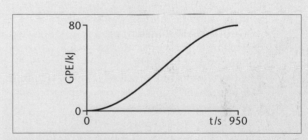

Fig A.5 *GPE increases most rapidly as vehicle passes point B*

Sketch graph (Figure A.5) showing:
Labelled axes and any line showing GPE increasing with time ✓
Sinusoidal shape ✓
(950 s, 80 kJ) labelled ✓
Either
no because passenger on other side is losing GPE ✓
if wheel is equally loaded/wheel balanced with people ✓
Or
yes   because no other passengers ✓
           so wheel is unequally loaded ✓

Part 2 Radioactivity

Quick test

1 D   Deflections of the alpha particles showed that an atom has a very small, central nucleus, containing almost all the atom's mass
2 C   The proton number is an integer; it is the number of protons in the nucleus
The nucleon number is the nearest integer to the mass number

3 B A nucleus of $^{209}_{83}$Bi has 209 nucleons and 83 protons
Number of neutrons = 209 − 83 = 126
As an atom is neutral, number of electrons = number of protons = 83

4 C A nucleus of $^{4}_{2}$He has 4 nucleons and 2 protons
Number of neutrons = 4 − 2 = 2
As an atom is neutral, number of electrons = number of protons = 2

5 C Protons and neutrons (nucleons) each have a structure of 3 quarks

6 B Paper would reduce the intensity of any alpha radiation
Some gamma radiation would penetrate 1 cm of aluminium

7 D Background count is 0.1 Bq
Increased count comes from source within the lead box
Only gamma radiation can pass through lead

8 C Alpha particle ($^{4}_{2}\alpha$) consists of 4 nucleons and 2 protons
Number of nucleons in product
nucleus = 220 − 4 = 216
Number of protons in product nucleus = 86 − 2 = 84

9 A Beta-minus decay occurs when a neutron in the nucleus decays into a proton so Y has one less neutron and one more proton than X
Also as they have different proton numbers (not B) they cannot be isotopes (not C)
The number of neutrons decreases by one while the number of protons increases by one so the number of nucleons (neutrons + protons) stays the same (not D)

10 B Gamma has no effect on the particle composition of a nucleus so number of protons is unchanged = 27

11 C The number of nucleons stays the same while the number of protons decreases by one
A proton in the nucleus has changed into a neutron so released particle is $^{0}_{+1}$e, a beta-plus particle

12 B Total number of nucleons in products is 7 + 4(from alpha) = 11
Total number of protons in products is 3 + 2(from alpha) = 5
Other particle must contain one (11 − 10) nucleon and no (5 − 5) protons so it is a neutron

13 C When emitting an alpha particle:
proton number decreases by two (to $Z − 2$)
nucleon number decreases by four (to $A − 4$)
If this nucleus then emits a beta-minus particle:
proton number increases by one (to $Z − 2 + 1 = Z − 1$)
nucleon number stays the same ($A − 4$)

14 C For an isotope, the nucleon number must change while the proton number stays the same
An alpha particle must be released to change the nucleon number (decreases it by 4)
This also reduces the proton number by two
so two beta-minus particles must be released to increase the proton number back to its original value

15 A The radon is decaying into a different element

16 A 40 days = 4 × 10 days = 4 half-lives
Fraction remaining = $\frac{1}{2} \times \frac{1}{2} \times \frac{1}{2} \times \frac{1}{2} = \frac{1}{16}$
48 g × $\frac{1}{16}$ = 3.0 g

17 A 16 years is very much shorter than the half-life of 1600 years
so activity will hardly change for this source over this period of time

18 D 100 s = 2 half-lives so fraction remaining = $\frac{1}{2} \times \frac{1}{2} = \frac{1}{4}$
150 s = 3 half-lives so fraction remaining = $\frac{1}{2} \times \frac{1}{2} \times \frac{1}{2}$
$= \frac{1}{8}$
200 s = 4 half-lives so fraction remaining = $\frac{1}{2} \times \frac{1}{2} \times \frac{1}{2} \times \frac{1}{2}$
$= \frac{1}{16}$
so about $\frac{1}{2}$ g of undecayed radon will be left after 200 s

19 C Only gamma radiation will penetrate the pipe and 0.4 m of earth

Only want the isotope to remain in the water for long enough for the pipe to be detected

20 C Activity = λN
N = activity/λ = 7560 Bq/(0.0126 s^{-1}) = 600 000

Practice questions

1 (majority of) alpha particles pass straight through/are not deflected ✓
(some) alpha particles are deflected slightly ✓
(a small number of) alpha particles bounce back ✓

Majority pass straight through so atom must be mainly space ✓
Deflections produced by electrostatic forces, so nucleus must be charged ✓
Small number bounce back so charge must be very concentrated ✓

(i) range 10^{-11} m to 10^{-9} m ✓
(ii) range 10^{-15} m to 10^{-13} m ✓

2

	Alpha particle scattering	Deep inelastic scattering
Incident particles	Alpha particles	Electrons/neutrinos/ leptons ✓
Target	Gold foil/ atoms/nuclei ✓	Nucleons

Alpha particle scattering:
Atom is mainly space/atom contains a very small nucleus ✓
Atom contains a dense/massive nucleus ✓

Deep inelastic scattering:
Nucleons/protons/neutrons have a substructure/are not fundamental particles ✓
of quarks (only if previous mark scored) ✓

3 Source not present/source well away from GM tube (>1 m) ✓

Determine count over at least 1 minute/determine a number of count rates and average ✓

Clarity mark – answer needs to make sense on first reading and attempt to address the question asked ✓
to demonstrate no alpha:
GM moved away from very close to source; no sudden drop in count rate ✓
Shows no alpha since alpha stopped by a few centimetres of air ✓
to demonstrate no gamma:
Aluminium put between GM tube and source; reading decreases to almost background ✓
Shows no gamma since gamma not stopped by aluminium ✓

4 Number of neutrons = 234 – 92 = 142
Uranium correctly marked at (92, 142) on grid (Figure A.6) ✓

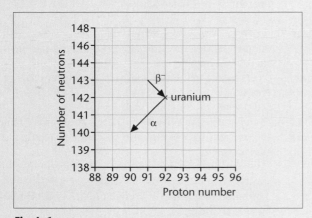

Fig A.6

Beta-minus decay marked from (91, 143) to (92, 142) ✓
Alpha decay marked with proton number decreasing from 92 to 90 ✓
and neutron number decreasing from 142 to 140 ✓

5 Graph has levelled off above zero (can't be background as already subtracted) ✓

Activity at which graph levels off = 150 cpm ✓

Attempt to allow for 150 cpm ✓

Correct use of graph to find an average half-life ✓
Half-life = 1.2 min ✓
e.g.
initial activity of M = (3000 – 150) cpm = 2850 cpm
This will fall to 1425 cpm after one half-life giving a total activity of (1425 + 150) cpm = 1575 cpm
Graph shows this activity is reached after 1.2 min
Activity of M after
1.8 min = (1150 – 150) cpm = 1000 cpm

This will fall to 500 cpm after one half-life giving a total activity of (500 + 150) cpm = 650 cpm
Graph shows this activity is reached after 3.0 min, giving a half-life of (3.0 – 1.8) min = 1.2 min

$$^1_0 n \rightarrow \, ^1_1 p + \, ^{\,0}_{-1}\beta \quad ✓$$

Wait until the graph has levelled off ✓
If activity is reduced to zero by paper then it is only alpha ✓
If not effected by paper but reduced to zero by aluminium then it is beta ✓
If not effected by paper or aluminium but reduced by lead then it is gamma ✓ **Max 3**

6 Number of protons = 6 ✓
Number of neutrons = 14 – 6 = 8 ✓

Volume of sphere = $4\pi r^3/3 = 4\pi \times (2.70 \times 10^{-15} \text{ m})^3/3$ ✓
$= 8.2 \times 10^{-44} \text{ m}^3$ ✓

Density = $m/V = 2.34 \times 10^{-26} \text{ kg}/(8.2 \times 10^{-44} \text{ m}^3)$ ✓
$= 2.8 \times 10^{17} \text{ kg m}^{-3}$ ✓

Nuclear density is very much greater than the densities of everyday materials ✓

Average ✓
Time taken for half the nuclei of that radioactive element to decay ✓

$\lambda t_{\frac{1}{2}} = 0.69$
$\lambda = 0.69/t_{\frac{1}{2}} = 0.69/[(5700 \times 365.25 \times 24 \times 60 \times 60) \text{ s}]$ ✓
$= 3.8 \times 10^{-12} \text{ s}^{-1}$ ✓

7 any two from:
radioactive rocks/radon gas/cosmic rays (or cosmic wind)/fallout/leaks from nuclear installations/named materials e.g. uranium, granite, carbon-14 ✓✓

$$^{22}_{11}\text{Na} \rightarrow \, ^{22}_{10}\text{Ne} + \, ^{\,0}_{1}\beta^+ \; (or \, ^{\,0}_{1}\text{e}) \; ✓$$

any one
from $^{20}\text{Na} \quad ^{21}\text{Na} \quad ^{23}\text{Na} \quad ^{24}\text{Na} \quad ^{25}\text{Na} \quad ^{26}\text{Na}$ ✓

$\lambda t_{\frac{1}{2}} = 0.69$
$\lambda = 0.69/t_{\frac{1}{2}} = 0.69/[(2.6 \times 365.25 \times 24 \times 60 \times 60) \text{ s}]$ ✓
$= 8.4 \times 10^{-9} \text{ s}^{-1}$ ✓

$A = \lambda N$
$2.5 \text{ Bq} = 8.4 \times 10^{-9} \text{ s}^{-1} \times N$ ✓
$N = 2.5 \text{ Bq}/(8.4 \times 10^{-9} \text{ s}^{-1}) = 3.0 \times 10^8$ ✓

Salt is not heavily contaminated
since this is a small number of atoms (compared to number of atoms in a spoonful of salt) ✓
or since its count rate is less than background

Unit 2 Electricity and thermal physics

Part 1 Electricity

Quick test

1 A Rate of flow of charge = current
Current is measured in amperes

2 C During a single revolution, a total charge of $4q$ passes any given point alongside the rim
Since disk makes n revolutions each second, total charge passing = $4q \times n$
$I = Q/t = 4qn/(1 \text{ s}) = 4qn$

3 D $I = nAqv$
So $I \propto q$; directly proportional rather than inversely proportional

4 B In a metal, n remains constant
The increased potential difference forces the charge carriers move faster along the wire

5 B Voltmeter A reads 6 V
Voltmeter B reads 6 V ÷ 2 = 3 V
Voltmeter C reads 6 V
Voltmeter D reads 6 V ÷ 3 = 2 V

6 C $Q = It = 2 \text{ A} \times 5 \text{ s} = 10 \text{ C}$ ... correct
Potential difference = 15 V = 15 J C^{-1}... correct
Energy dissipated = 15 J C^{-1} × 10 C = 150 J (or VIt = 150 J) ... incorrect
$R = V/I = 15 \text{ V}/(2 \text{ A}) = 7.5 \text{ }\Omega$ ... correct

7 D 3.3 Ω connect all 3 resistors in parallel ($R = 10 \text{ }\Omega \div 3$)
6.7 Ω connect 2 in series with the other in parallel $[1/R = 1/(20 \text{ }\Omega) + 1/(10 \text{ }\Omega)]$
15 Ω connect 2 in parallel with the other in series $[R = (10 \text{ }\Omega \div 2) + 10 \text{ }\Omega]$
20 Ω can only be obtained by using 2 of the resistors in series

8 C Jane's voltmeter reads the total potential difference across both the resistor and the ammeter
Margaret's voltmeter reads only the potential difference across the resistor
Jane's ammeter reads only the current in the resistor
Margaret's ammeter reads the total current in both the resistor and the voltmeter

9 D Drift speed will be lower
Length of the conductor will not increase significantly
Number of free electrons will remain the same

10 C Graph shows resistance decreasing as temperature increases

11 A Area = 0.5 mm² = 0.5×10^{-6} m²
$R = \rho l/A$
$\rho = RA/l = 3 \text{ }\Omega \times 0.5 \times 10^{-6} \text{ m}^2/(10 \text{ m}) = 1.5 \times 10^{-7} \text{ }\Omega \text{ m}$

12 D Rearrange each equation to give expression for k
Correct relationship is that which gives approximately the same value of k for all four wires
i.e. $k = R \times d^2 = 32 \times 0.5^2 = 8.0 \times 1.0^2$
$= 3.6 \times 1.5^2 = 2.0 \times 2.0^2 \approx 8$

13 B Current I = battery e.m.f./total resistance = 12 V/(15 Ω + 45 Ω) = 0.2 A
Potential difference across 15 Ω resistor = IR = 0.2 A × 15 Ω = 3 V
Potential difference across 45 Ω resistor = 0.2 A × 45 Ω = 9 V

14 A Circuit A: diode is in reverse bias and so does not conduct
Circuits B and D: diode will conduct for the half cycle of a.c. for which it is in forward bias
Circuit C: diode is in forward bias and so conducts

15 C With diode connected as shown:
Diode conducts so circuit is effectively a 2 Ω resistor in series with two 2 Ω resistors in parallel
Resistance of parallel combination = 1 Ω
Potential difference across parallel combination = $\frac{1}{2}$ × potential difference across 2 Ω = 6 V
Potential difference of supply = 12 V + 6 V = 18 V
When diode is reversed:
diode does not conduct so circuit is effectively two 2 Ω resistors in series
potential difference across each resistor = $\frac{1}{2}$ × 18 V = 9 V

16 D Power $P = IV = I^2R = V^2/R$
Both resistors have the same potential difference across them
6 Ω resistor will have 1.5 times the current as 9 Ω resistor
so power will be 1.5 × 36 Ω = 54 Ω

17 B e.m.f. is the energy given to each coulomb of charge

18 A No current in broken lamp so ammeter reading decreases
Less current through internal resistance so less 'lost volts'
Terminal potential difference of battery is more and lamps are brighter

19 B When switch is closed:
External circuit is two 6 Ω resistors in parallel giving a combined resistance of 3 Ω
Total circuit resistance = 3 Ω + 2 Ω = 5 Ω
Circuit current = 12 V/(5 Ω) = 2.4 A
Current in each branch = $\frac{1}{2}$ × 2.4 A = 1.2 A

20 C Potential difference across 60 Ω resistor = 30 V – 5 V = 25 V
Potential difference across PQ = 1/5 of potential difference across 60 Ω resistor
'resistance of parallel combination = 60 Ω/5 = 12 Ω
$1/(12 \text{ }\Omega) = 1/(48 \text{ }\Omega) + 1/R$
$1/R = 1/(12 \text{ }\Omega) - 1/(48 \text{ }\Omega) = 0.0625 \text{ }\Omega^{-1}$
$R = 1/(0.0625 \text{ }\Omega^{-1}) = 16 \text{ }\Omega$

Practice questions

1 n: number of charge carriers per unit volume ✓
v: drift (average) velocity of charge carriers ✓

Unit of n is m^{-3} ✓
Unit of A is m^2, Q is C, v is m s^{-1} ✓
Unit of $nAQv$ = $\text{m}^{-3} \times \text{m}^2 \times \text{C} \times \text{m s}^{-1}$
$= \text{C s}^{-1} = \text{A}$ = unit of I ✓

Value of n is much larger for a metal than an insulator ✓
so current is much larger in a metal ✓

2 AB: uniform *or* constant acceleration ✓
BC: sudden deceleration ✓

AB: power supply forces the electron to accelerate ✓
BC: electron collides with a lattice ion ✓

Drift velocity is the average rate of progress of the electron – its average velocity ✓
Horizontal line added to graph in Figure 2.30 at approximately $\frac{1}{2}v_{max}$ ✓

Electron's collision makes lattice ion vibrate more vigorously *or* electron's kinetic energy is transferred to the lattice ✓

3 $P = V^2/R$ ✓
$R = V^2/P = (230\ \text{V})^2/(100\ \text{W})$ ✓
$= 529\ \Omega$ ✓

Current same throughout series circuit ✓
Reference to $I = nAQv$ ✓
Value of n similar for all metals in circuit ✓
Filament has smaller A than connecting wires ✓
So v has to be higher to give the same I ✓ **Max 4**

4 $V = IR$ ✓
$R = V/I = 1.5\ \text{V}/(100\ \mu\text{A})$ ✓
$= 15\ \text{k}\Omega$ ✓

(a) to half current must double total resistance so need to add another 15 kΩ ✓
(b) to quarter current must quadruple total resistance so need to add another 45 kΩ ✓

Resistance values added to scale in Figure 2.31 with:
0 kΩ aligned with 100 μA ✓
15 kΩ aligned with 50 μA ✓
45 kΩ aligned with 25 μA ✓

5 Use equation $\rho = RA/l$ ✓
Where $A = tw$ ✓

$l = RA/\rho = Rtw/\rho$ ✓
$= 1.0\ \Omega \times 0.15 \times 10^{-3}\ \text{m} \times 20 \times 10^{-3}\ \text{m}/(2.7 \times 10^{-8}\ \Omega\ \text{m})$ ✓
$= 111\ \text{m}$ ✓

Reduce width w of strip *or* use thinner foil/reduce t ✓
Measurement of smaller sizes will increase error in experiment ✓

6 No because the graph is not a straight line through the origin ✓

From graph, when $V = 0.74$ V current $I = 80$ mA
(78 – 82 mA allowed) ✓
$R = V/I = 0.74\ \text{V}/(80 \times 10^{-3}\ \text{A}) = 9.25\ \text{V}$ ✓

Potential difference across $R = 9.0\ \text{V} - 0.74\ \text{V} = 8.26\ \text{V}$ ✓
$R = V/I = 8.26\ \text{V}/(80 \times 10^{-3}\ \text{A})$ ✓
$= 103\ \Omega$ (100 to 106 Ω allowed) ✓

Vertical line added to grid in Figure 2.33 ✓
at $V = 0.7$ V ✓

7 **(i)** power dissipated in external resistance $= I^2R$ ✓
(ii) power dissipated in internal resistance $= I^2r$ ✓
(iii) rate of conversion of chemical energy in cell $= \varepsilon I$ ✓

Total energy in = total energy out
$\varepsilon I = I^2R + I^2r$ ✓
so, by cancelling I, $\varepsilon = IR + Ir = I(R + r)$ and $I = \varepsilon/(R + r)$ ✓

Maximum current occurs when R is zero ✓
$I_{max} = \varepsilon/r$ *or* larger r means smaller I ✓

1 MΩ ✓
as this would give the smallest maximum current ✓

8 Circuit diagram (Figure A.7) of torch showing:
a circuit in which the lamp will light ✓
three cells and a lamp all in series (no resistances unless labelled as internal) ✓

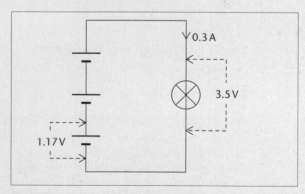

Fig A.7

Labels added to circuit diagram:
3.5 V shown across the lamp ✓
arrow labelled 0.3 A on lead to or from lamp ✓
1.17 V shown across at least one of the cells *or* 3.5 V across all three ✓

Total 'lost volts' $= 4.5\ \text{V} - 3.5\ \text{V} = 1.0\ \text{V}$
Total internal resistance $= 1.0\ \text{V}/(0.3\ \text{A}) = 3.3\ \Omega$
Internal resistance of one cell $= 3.3\ \Omega/3 = 1.1\ \Omega$
or
total resistance $= 4.5\ \text{V}/(0.3\ \text{A}) = 15\ \Omega$ ✓
lamp resistance $= 3.5\ \text{V}/(0.3\ \text{A}) = 11.7\ \Omega$ ✓
internal resistance $= (15 - 11.7\ \Omega)/3 = 1.1\ \Omega$ ✓

Part 2 Thermal physics

Quick test

1 C kg m^{-3} = mass/volume so this is the unit of density
J m^{-3} = N m m^{-3} = N m^{-2} = Pa are all possible units of pressure

2 D $p = F/A = mg/A$ = 50 kg × 9.81 N kg^{-1}/(0.5 m^2) = 981 N

3 B T_1 = (77 + 273) K = 350 K and T_2 = (27 + 273) K = 300 K
Using $p_1/T_1 = p_2/T_2$
$p_2 = p_1T_2/T_1$ = 210 kPa × 300 K/(350 K) = 180 kPa

4 A From Boyle's law, pressure is inversely proportional to volume
$p \propto 1/V$ so that pV = constant

5 D Volume increases as pressure decreases
Pressure increases with depth in water

6 C You must learn this equation and the meaning of all its symbols

7 B The collisions with the wall are the important factor in determining the pressure
the change of momentum (= $2mv$) is significant

8 C A quick look at the provided data (Appendix 2) will confirm this!
$<c^2>$ is the mean square speed, the square of the mean speed would be $<c>^2$
$\frac{1}{2}m<c^2>$ is energy so units of k must be J K^{-1} for equation to be homogeneous
The more gas molecules, the greater the total mass and the greater the density ρ

9 A Using $p = (1/3)\rho<c^2>$
$<c^2> = 3p/\rho$ = 3 × 1.0 × 10^5 Pa/(1.2 kg m^{-3})
= 250 000 m^2 s^{-2}
Root mean square
speed = $\sqrt{<c^2>}$ = $\sqrt{(250\ 000\ \text{m}^2\ \text{s}^{-2})}$ = 500 m s^{-1}

10 B Sum of square speeds = (1^2 + 2^2 + 3^2 + 4^2 + 5^2) km^2 s^{-2}
= 55 km^2 s^{-2}
Mean square speed = 55 km^2 s^{-2}/number of molecules
= 11 km^2 s^{-2}
Root mean square speed = $\sqrt{(11\ \text{km}^2\ \text{s}^{-2})}$ = $\sqrt{11}$ km s^{-1}

11 A Reducing the temperature decreases the mean speed of the molecules (peak shifts to the left)
but increases the proportion of molecules with that speed (peak rises)

12 D Forces between molecules in an ideal gas are negligible so there is no molecular potential energy
Internal energy of ideal gas = total molecular kinetic energy
Molecular kinetic energy depends on temperature only

13 C Energy supplied = $mc\Delta\theta$
Mass of bath water is much larger than the other masses
although the others have greater temperature rises, their masses are much smaller

14 B Energy
required = $mc\Delta\theta$ = 2 kg × 4200 J kg^{-1} K^{-1} × (100 °C – 20 °C) = 672 000 J
Energy supplied = power × time
t = 672 000 J/power = 672 000 J/(1000 W)
= 672 s = 11.2 min

15 C Total kinetic energy = 2 × ($\frac{1}{2}mv^2$) = mv^2 = 90 000 m
since they join, mass being warmed up is $2m$
$2mc\Delta\theta$ = 90 000 m
$\Delta\theta$ = 90 000/2c = 90 000/(2 × 500) = 90 °C

16 D Specific heat capacity would be calculated using
$c = VIt/m\Delta\theta$
Any energy losses to surroundings will reduce $\Delta\theta$ so c would be more
if current I is too big then c will also be too big
Any energy absorbed by beaker will reduce $\Delta\theta$ so c would be more
if potential difference is too small then c will also be too small

17 A 1 kg of substance takes 3 min and 6000 J (3 × 2000) to melt so its latent heat of fusion is 6000 J kg^{-1}
After 4 min, the substance starts to change into a liquid
Graph shows the solid warms up slower than the liquid so solid has the greater specific heat capacity
The substance must be pure as it has a precise melting point

18 B Temperature of lamp will be constant so no change in its internal energy
Electrical work
done = power × time = 60 W × (30 × 60) s = 108 000 J
Hot lamp is heating colder surroundings so ΔQ is negative
Equation still applies, in this case
0 = –108 000 J + 108 000 J

19 B T_1 = 550 °C = 823 K
Using maximum thermal efficiency = 1 – T_2/T_1
$T_2 = T_1$ × (1 – maximum thermal efficiency) = 823 K × (1 – 0.45) = 453 K = 180 °C

20 D Energy has to be removed from inside the fridge at same rate (30 W) as it enters
but heat pump has to do work to extract the energy
so total rate of energy release into kitchen is more than 30 W

Practice questions

1 Clarity mark – answer needs to make sense on first reading and attempt to address the question asked ✓
Pressure equal throughout the hydraulic fluid ✓
Since $p = F/A$, $F = pA$ ✓
So larger area gives a larger force ✓

2 Using $p_1/T_1 = p_2/T_2$ ✓
$p_2 = p_1T_2/T_1$
= 1.00 × 10^5 Pa × (100 + 273) K/[(0 + 273) K] ✓
= 1.37 × 10^5 Pa ✓

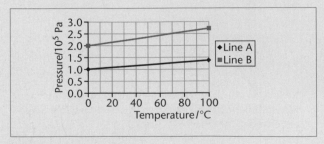

Fig A.8

Accurate graph (Figure A.8) showing line A:
Positive gradient straight line ✓
from 100 kPa at 0 °C to 137 kPa at 100 °C ✓

Line B on same graph:
Positive gradient straight line above line A for all its length ✓
from 200 kPa at 0 °C to 273 kPa at 100 °C ✓
(since volume is halved, pressure is doubled)

3 Sketch graph (Figure A.9) showing:

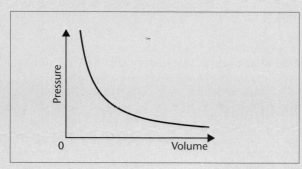

Fig A.9

Pressure decreasing as volume increases ✓
Smooth curve, asymptotic to both axes ✓

As volume increases the molecules get further apart ✓
so collide less often with the walls of their container ✓

Use pressure = force/area ✓
           = weight of known mass/cross-sectional area
                 of top of piston ✓
and add this to the atmospheric pressure ✓

Examiner's comments

It is easy to forget that atmospheric pressure is always acting!

e.g. temperature not constant ✓
      leakage of air from syringe ✓
      friction between piston and cylinder ✓
      weight of piston not included in pressure
      calculation ✓ **Max 1**

4 Large volume increases when liquid changes to gas/gases
are easy to compress ✓
implies large spacing of gas molecules ✓

Brownian motion of smoke particles in a gas ✓
smoke particles randomly knocked about by moving gas particles ✓

5 Rate of rise of temperature is gradient of tangent to graph
initial gradient = $\Delta\theta/\Delta t$ = (25 °C – 0 °C)/(12 min) ✓
           = 25 K/[(12 × 60) s] = 0.035 K s^{-1}
               (range: 0.030 K s^{-1} to 0.042 K s^{-1}) ✓

Examiner's comments

Since $\Delta\theta$ is a change in temperature, its value is the same in both °C and K.

Energy transfer = $mc\Delta\theta$
Rate of energy transfer = $mc\Delta\theta/\Delta t = mc \times$ gradient ✓
  = 400 × 10^{-3} kg × 4200 J kg^{-1} K^{-1} × 0.035 K s^{-1} ✓
  = 59 W (range: 50 W to 71 W) ✓

During the initial 27 min the ice is melting ✓
Melting occurs at a constant temperature ✓

Energy = power × time = 59 W × (27 × 60) s ✓
  = 95 580 J ✓
Using energy transfer = $l\Delta m$
Δm – 95 580 J/(330 kJ kg^{-1}) ✓
  = 0.29 kg (range: 0.24 to 0.35 kg) ✓

6 Between P and Q the ice is melting:
Regular arrangement of the molecules breaks
down/arrangement becomes disordered ✓
as the bonds between the molecules weaken ✓
Molecules start to move relative to each
other ✓ **Max 2**

Time taken to melt = 39 min – 2.5 min = 36.5 min
  – (36.5 × 60) s = 2190 s
Energy absorbed = power × time = 30 W × 2190 s
  = 65 700 J ✓
Since energy needed = $l\Delta m$
l = energy needed/Δm = 65 700 J/(0.20 kg) ✓
  = 3.3 × 10^5 J kg^{-1} ✓

Would need a thermometer and a clock/a temperature
sensor and a datalogger ✓
record temperature at regular time intervals ✓

e.g. insulate the beaker ✓
      stir to ensure even heating ✓
      measure temperature well away from heater ✓
      read thermometer at eye-level to avoid parallax
      errors ✓ **Max 1**

Line added to graph:
with a horizontal section at 29 °C (melting point) ✓
Both gradients steeper than for water (smaller specific
heat capacity so it warms up quicker) ✓
Shorter horizontal section (smaller specific latent heat so
less energy required to melt) ✓

7 $\Delta U = 0$ ✓
since temperature remains constant ✓

$\Delta W = 600$ J ✓
electrical work done = power × time
           = 60 W × 10 s = 600 J ✓

$\Delta Q = -600$ J ✓
$\Delta Q = \Delta U - \Delta W = 0$ J – 600 J = –600 J ✓

8 A device that takes energy from a hot source ✓
converts some of this energy into useful work ✓
and transmits the rest of it to a cold sink ✓

A higher working temperature gives a higher
efficiency ✓

Unit 3 Topics and practical test

Part 1 Topics

Astrophysics

(a) Power is the rate of doing work or power = work
done/time ✓
unit = W or J s^{-1} ✓
Base units: J = N m and N = kg m s^{-2} ✓
W = N m s^{-1} = kg m s^{-2} m s^{-1} = kg m^2 s^{-3} ✓

(b) Sun marked on main sequence where $L_\odot = 10^0 = 1$ ✓
Massive main sequence star marked on main sequence at top
left ✓
Temperature scale added showing increasing temperatures to
the left ✓
coolest = 3000 (±1000) and hottest = 35 000 (±15 000) ✓

Large mass stars are brighter/have greater luminosity/are
hotter ✓
so burn up fuel/hydrogen quickly ✓

$\Delta E = c^2 \Delta m$ ✓
so luminosity = $\Delta E/\Delta t = c^2 \Delta m/\Delta t$
$\Delta m/\Delta t$ = luminosity/c^2 = 3.9 × 10^{26} W/(3 × 10^8 m s^{-1})^2 ✓
     = 4.3 × 10^9 kg s^{-1} ✓

(c) Can detect fainter stars/more distant stars/very small
amounts of light ✓
Takes less time to get an image or more images per
session ✓
Can detect wavelengths that would be otherwise absorbed by
atmosphere ✓
Less dust/scattering/light pollution ✓
No refraction/twinkling ✓ **Max 2**

(d) A neutron star ✓
Signals are regular/at precise intervals ✓
Pulses are short/fast ✓

Clarity mark – answer needs to make sense on first reading
and attempt to address the question asked ✓
Not continuous as star spins/rotates ✓
Radio beam passes Earth as it sweeps round ✓

Further detail e.g. diagram/narrow beam/lighthouse
analogy ✓ **Max 3**
(e) Red giants: cool high volume/big/bright stars ✓
Parallax displacement: compares position of a star relative to
a distant star ✓
with that viewed 6 months later ✓

Period = 5 (± $\frac{1}{2}$) days ✓
Luminosity of B must be greater (since its period is
longer) ✓
B must be further away ✓
e.g. since $I = L/4\pi D^2$ so larger L needs larger D to give
same I ✓

Gravitational force causes a star to contract ✓
Forces due to radiation pressure causes a star to expand ✓

Solid materials

(a) Work = force × displacement/distance moved in force
direction ✓
unit = J or N m ✓
base units: J = N m and N = kg m s^{-2} ✓
J = kg m s^{-2} m = kg m^2 s^{-2} ✓

(b) Extension is proportional to force/load ✓
below the elastic limit ✓
Ultimate tensile stress = maximum = 2.3 × 10^8 Pa ✓
Young modulus = slope of linear region
     = 2.1 × 10^8 Pa/(1.6 × 10^{-3}) ✓
     = 1.3 × 10^{11} Pa ✓

Stress = F/A = 250 N/(1.7 × 10^{-6} m^2) = 1.5 × 10^8 Pa ✓
so elastic as on linear part of graph ✓
Point P plotted at correct stress ✓
From graph when stress = 1.5 × 10^8 Pa, strain = 1.1 × 10^{-3} ✓
extension = l × strain = 3.0 m × 1.1 × 10^{-3} = 3.3 × 10^{-3} m ✓

(c) Suitable diagram e.g. Figure A.10 ✓

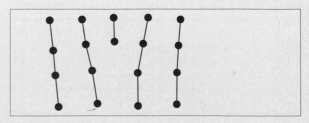

Fig A.10

Dislocation is an extra/missing half row/plane of
atoms/ions ✓

Clarity mark – answer needs to make sense on first reading
and attempt to address the question asked ✓
Dislocation moves ✓
and blunts tip of crack ✓
So reduces stress ✓ **Max 3**

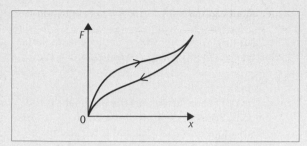

Fig A.11

(d) Graph similar to that in Figure A.11
    axes and shape ✓
    arrowheads/labels to show loading and unloading ✓
Area (between the two graphs) represents energy ✓
converted to internal energy (in rubber band) ✓

(e) Composite material: two (or more) materials bonded/joined/combined ✓
to make use of the (best) properties of both ✓
A named example other than dentine ✓ **Max 2**
Diagram (Figure 7.15) of crystalline material with atoms/planes of atoms labelled ✓✓

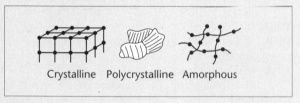

Crystalline   Polycrystalline   Amorphous

Fig A.12

Diagram (Figure A.12) of amorphous polymer with strands/molecules labelled ✓
(i) Tension line continued back down into lower left quadrant ✓
    to a greater stress ✓
(ii) Line added to upper right quadrant with a smaller slope ✓
    and a larger area under it ✓

Nuclear and particle physics

(a) Density = mass per unit volume ✓
Volume ratio = $(10^{-10})^3/(10^{-15})^3 = 10^{-30}/10^{-45}$ ($= 10^{15}$) ✓
Density of gold
nucleus = $19\,000$ kg m$^{-3} \times 10^{15} = 1.9 \times 10^{19}$ kg m^{-3} ✓
assuming mass of nucleus = mass of atom or electrons have negligible/zero mass ✓

(b) Energy spectra (Figure A.13) for typical α and β^- decays

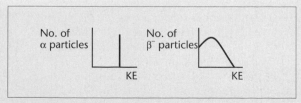

Fig A.13

α: vertical line or narrow peak ✓
β^-: correct shape ✓
and shows upper limit to k.e. (intercepts axis) ✓
$n \rightarrow p + e^- + \bar{\nu}$ ✓✓
n = (udd) and p = (uud) ✓
Clarity mark – answer needs to make sense on first reading and attempt to address the question asked ✓
decay involves a down quark changing to an up quark ✓
Change of quark type means it can only be the weak interaction ✓ **Max 3**

(c) For strontium-90, N = 90 – 38 = 52 so position marked at (38,52) ✓
β^-: neutron $\rightarrow$ proton so N decreases by 1 and Z increases by 1; yttrium marked at (39,51) ✓
N = 82 – 37 = 45 so rubidium marked at (37, 45) ✓
β^+ (positron) or α decay ✓

(d) Charge on u = +2/3 so charge on $\bar{u}$ = –2/3 ✓
Total charge on K$^-$ = –1 so charge on s = –1/3 ✓
Using charge conservation, (–1) + (+1) $\rightarrow$ (0) + X ✓
X must be neutral ✓
Using baryon number conservation, (0) + (+1) $\rightarrow$ (+1) + X ✓
X is a meson ✓
X = (s$\bar{d}$) ✓
Must be s since strong interaction and quark type unchanged or strangeness conserved ✓
must be $\bar{d}$ for X to be neutral and a meson, X(0) = s(–1/3) + $\bar{d}$(+1/3) ✓

(e) positron, (charmed) antiquark, π^- ✓ for 2 ✓✓ for 3
$\Psi = c(+1) + \bar{c}(-1)$ so no overall charm ✓
By moving around each other in high energy orbitals ✓
particle A = Ψ ✓
particles B and C = π^+ and π^- ✓
the strong interaction ✓
gluon ✓

Medical physics

(a) Potential difference – work done per unit charge or power per unit current ✓
unit = V or J C^{-1} ✓
Base units: J = N m and N = kg m s^{-2} ✓
C = A s so
J C^{-1} = N m A^{-1} s^{-1} = kg m s^{-2} m A^{-1} s^{-1} = kg m^2 A^{-1} s^{-3} ✓

(b) Some of the isotope removed by biological processes e.g. excretion ✓
$1/t_e = 1/t_r + 1/t_b = 1/(8 \text{ days}) + 1/(21 \text{ days})$ ✓
$t_e = 5.8$ days ✓
$1/8 = (\frac{1}{2})^3 = 3$ half-lives ✓
Time = 3×5.8 days = 17 days ✓
Gamma can be detected outside body/weakly ionising/does not interact with tissue ✓
Must be taken up by organ under investigation ✓

(c) Clarity mark – answer needs to make sense on first reading and attempt to address the question asked ✓
Z_{air} and Z_{body} very different so ultrasound strongly reflected ✓
Little ultrasound reaches organs being investigated ✓

Coupling medium excludes air between transmitter and skin ✓ **Max 3**
Coupling medium: gel/water/Vaseline ✓
Time delay = 2.7 div × 50 μs div^{-1} = 135 μs ✓
Distance = speed × time = $1.5 × 10^3$ m s^{-1} × 135 × 10^{-6} s ✓
    = 0.20 m ✓
as this is 'there and back', head = $\frac{1}{2}$ × 0.20 m = 0.10 m ✓

(d) Since diagnostic, operating voltage in kV range ✓
Rotated to prevent one point on anode being overheated ✓
Evacuated so no atoms for electrons to collide with/scatter off ✓
Outer case made from lead ✓

(e) Heterogeneous: containing X-rays of many wavelengths/frequencies/energies ✓
Hardening: removing longer wavelength/lower energy X-rays ✓

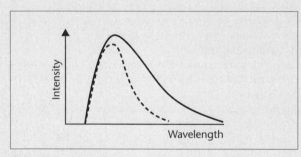

Fig A.4

Graph (Figure A.4) showing:
    less area enclosed ✓
    (almost) the same at small λ ✓
    (much) lower at long λ ✓
More penetrating as remaining X-rays have a higher average energy/shorter average wavelength ✓
since long λ attenuated more than short λ/the beam is harder/the beam is of higher quality ✓
Beneficial: low energy X-rays, which are of no use, are absorbed by the patient ✓
so filtering these out reduces the dose received ✓

Unit 4 Waves and our universe

Part 1 Oscillations and waves

Quick test

1 B   There are 2π radians in a circle so cog makes 3 revolutions each second
number of teeth hitting strip each second = 3 × 25 = 75 Hz

2 C   Linear momentum is the only vector quantity so linear momentum changes as the body changes direction
3 C $a = v^2/r = (6.3 × 10^3$ m s$^{-1})^2/(10 × 10^6$ m) = 3.97 m s^{-2}
4 C   Apparent weightlessness occurs because the astronaut is falling freely
The only force acting on the astronaut is the gravitational attraction of the Earth
This supplies the centripetal force and makes the astronaut follow a circular path
5 A   Speed of the target is least at the extremes of its oscillation
so target spends more time in regions 1 and 5
6 C $T = 2π\sqrt{(l/g)}$
for Moon, $l/(g/6) = 6l/g$
$T_{Moon} = 2π\sqrt{(6l/g)} = \sqrt{6} × T_{Earth} = \sqrt{6}$ s
7 D   Period $T = 2$ s so frequency $f = 1/T = 0.5$ s^{-1}
Maximum acceleration
$a = (2πf)^2x_o = 4 × π^2 × 0.25$ s^{-2} × 4 m = $4π^2$ m s^{-2}
8 B   The engine is forcing the mirror to oscillate
When running slowly, engine frequency matches natural frequency of mirror
Mirror responds and vibrates with large amplitude so reflection blurs
9 B   Amplitude = maximum displacement = 2 mm
$f = 1/T = 1/(20 × 10^{-6}$ s) = $5 × 10^4$ Hz = 50 kHz
10 D
11 B   Sound is a longitudinal wave
Longitudinal waves cannot be polarised
12 A   Make sure you learn both the order of the electromagnetic spectrum and of the colours of visible light
13 B $λ = c/f = 3 × 10^8$ m s$^{-1}/(1 × 10^{15}$ Hz)
    = $3 × 10^{-7}$ m = 300 nm
This is slightly shorter than that of the violet end of the visible spectrum
14 D   Pulses move through each other and their displacements add up
Resultant displacement at some point must be bigger than initial displacement
D does not go above the initial displacement – can you draw what it should look like?
15 C   A bright fringe occurs when the path difference is zero
$s = λD/x$ (so s will only equal x in the unlikely situation where $λD = x^2$!)
The two slits act as coherent sources
16 A $x = λD/s$ so need to either increase λ or D, or decrease s
red light has a longer wavelength than green light so this increases λ
read B carefully, D is the distance between slits and screen, not source and slits
17 A $s = λD/x = 3 × 10^{-3}$ m × 2 m/(5 × 10^{-2} m) = 0.12 m
18 B   Longitudinal waves cannot be polarised
Transverse stationary waves can be polarised (as in microwave experiment)
19 D   Points either side of a node do vibrate out of phase but with a phase difference of π rad or 180°
20 C $λ = v/f = 75$ m s$^{-1}/(50$ Hz) = 1.5 m
Distance between adjacent nodes = $λ/2$ = 1.5 m/2 = 0.75 m

Practice questions

1 $\omega = 2\pi/T = 2\pi/[(365 \times 24 \times 60 \times 60) \text{ s}]$ ✓
$= 1.99 \times 10^{-7} \text{ rad s}^{-1}$ ✓

$F = mv^2/r = mr\omega^2$ ✓
$= 5.98 \times 10^{24} \text{ kg} \times 1.50 \times 10^{11} \text{ m}$
$\times (1.99 \times 10^{-7} \text{ rad s}^{-1})^2$ ✓
$= 3.56 \times 10^{22} \text{ N}$ ✓

Gravitational force of Sun attracting the Earth ✓

2 KE at lowest position = GPE at highest position ✓
$\frac{1}{2}mv^2 = mgh$
$v = \sqrt{2gh} = \sqrt{(2 \times 9.81 \text{ m s}^{-2} \times 0.8 \text{ m})}$ ✓
$= \sqrt{(15.7 \text{ m}^2 \text{ s}^{-2})} = 4.0 \text{ m s}^{-1}$ ✓

Resultant (centripetal)
force $= mv^2/r = 21 \text{ kg} \times (4.0 \text{ m s}^{-1})^2/(3.0 \text{ m})$ ✓
$= 110 \text{ N}$ ✓
Resultant force = force exerted up on child – weight of child
Force on child = resultant force + weight of child ✓
$= 110 \text{ N} + (21 \text{ kg} \times 9.81 \text{ N kg}^{-1}) = 110 \text{ N} + 206 \text{ N}$
$= 316 \text{ N}$ ✓

3 $F \propto -x$ or $F = -kx$ ✓
Towards the centre of the oscillation/towards the equilibrium position ✓
maximum ✓
maximum ✓

Using $T = 2\pi\sqrt{(m/k)}$ so $T^2 = 4\pi^2 m/k$
Force constant $k = 4\pi^2 m/T^2 = 4 \times \pi^2 \times 0.80 \text{ kg}/(1.5 \text{ s})^2$ ✓
$= 14 \text{ N m}^{-1}$ ✓

4 $T = 2\pi\sqrt{(l/g)} = 2 \times \pi \times \sqrt{[24.9 \text{ m}/(9.81 \text{ m s}^{-2})]}$ ✓
$= 10.0 \text{ s}$ ✓

$f = 1/T = 1/(10 \text{ s}) = 0.1 \text{ Hz}$
Maximum speed $= 2\pi f x_0 = 2 \times \pi \times 0.1 \text{ Hz} \times 3.25 \text{ m}$ ✓
$= 2.0 \text{ m s}^{-1}$ ✓

Maximum
acceleration $= (2\pi f)^2 x_0 = (2 \times \pi \times 0.1 \text{ Hz})^2 \times 3.25 \text{ m}$ ✓
$= 1.3 \text{ m s}^{-2}$ ✓

Two sketch graphs showing:
velocity shown as either sine or cosine ✓
acceleration either cosine or –sine, i.e. consistent with velocity ✓
Two complete cycles in 20 s ✓
Velocity and acceleration scales shown ✓

5 Steering wheel is being forced to vibrate ✓
It responds with maximum amplitude at one particular frequency ✓

when forcing frequency = natural frequency of steering wheel ✓
Maximum acceleration $= (2\pi f)^2 x_o$
$= (2 \times \pi \times 2.4 \text{ Hz})^2 \times 6.0 \times 10^{-3} \text{ m}$ ✓
$= 1.4 \text{ m s}^{-2}$ ✓

6 (a) using $I = P/4\pi r^2$
$r^2 = P/4\pi I$ ✓
$= 10 \times 10^3 \text{ W}/(4 \times \pi \times 2.2 \times 10^{-5} \text{ W m}^{-2})$ ✓
$r = \sqrt{(3.6 \times 10^7 \text{ m}^2)} = 6.0 \times 10^3 \text{ m}$ ✓

(b) distance between adjacent node and antinode $= \lambda/4$ ✓
$\lambda = 4 \times 0.8 \text{ m} = 3.2 \text{ m}$ ✓

$v = f\lambda = 95 \times 10^6 \text{ Hz} \times 3.2 \text{ m}$ ✓
$= 3.0 \times 10^8 \text{ m s}^{-1}$ ✓

This speed suggests that radiowaves are electromagnetic ✓

7 Clarity mark – answer needs to make sense on first reading and attempt to address the question asked ✓
Microwave transmitter ... 2 slits in metal barriers ... receiver ✓
Receiver connected to meter/loudspeaker ✓
Move receiver in arc centred on mid-point of slits ✓
Note the varying intensity/maxima and minimum ✓
Appropriate slit separation = 7 cm (range allowed = 4 to 10 cm) ✓

Examiner's comments

A good, labelled diagram would gain most of these marks.

Locate position of first maximum ✓
Measure distance from first maximum to each slit ✓
Difference gives wavelength ✓
(note: no marks here for using $\lambda = xs/D$ since s is not sufficiently small compared with D)

8 Using $\lambda = xs/D$
fringe spacing
$x = \lambda D/s = 690 \times 10^{-9} \text{ m} \times 3.5 \text{ m}/(0.50 \times 10^{-3} \text{ m})$ ✓
$= 4.8 \times 10^{-3} \text{ m}$ ✓
Sketch of pattern showing:
at least four equally spaced fringes ✓
shown as light and dark bands ✓

Fringe spacing decreases/new spacing = 3.2 mm ✓
Colour changes to blue/violet ✓

Central fringes ✓
and 3rd blue with 2nd red/6th with 4th/9th with 6th ✓

Part 2 Quantum physics and the expanding universe

Quick test

1 B $E = hf$ so $h = E/f$
units of $h = \text{J/s}^{-1} = \text{J s} = \text{N m s}$

2 C $E = hf = hc/\lambda$
$\lambda = hc/E = 6.63 \times 10^{-34}$ J s $\times 3.00 \times 10^8$ m s^{-1}/
$(3 \times 10^{-19}$ J$) = 6.63 \times 10^{-7}$ m
Wavelength of 663 nm is in the visible region

3 D Laser emits P joules every second
Photon energy $E = hc/\lambda$
Number of photons every second = total energy per
second/energy per photon $= P/(hc/\lambda) = P\lambda/hc$

4 C Greater intensity so more photons releasing more
photoelectrons
Same wavelength so same photon energy and same
maximum kinetic energy

5 D Kinetic energy $\propto$ speed^2
Maximum kinetic energy = photon energy – work
function $= hf$ – work function

6 D

7 A Work function = energy of a threshold frequency
photon $= hf_0$
Maximum kinetic energy $= hf - hf_0 = h(f - f_0)$
$= 6.63 \times 10^{-34}$ J s $\times (7.0 - 5.0) \times 10^{14}$ Hz
$= 1.33 \times 10^{-19}$ J

8 B Tungsten lamp uses a hot filament to produce a
continuous spectrum
Discharge lamp uses excitation and ionisation of a gas
to produce a spectrum containing only certain
frequencies linked to changes of energy states

9 A Photon energy $= hf$ = energy difference between two
levels $= E_1 - E_2$

10 B Photon energy $= hf = hc/\lambda = 2.2 \times 10^{-18}$ J
$\lambda = hc/(2.2 \times 10^{-18}$ J$)$
$= 6.63 \times 10^{-34}$ J s $\times 3.00 \times 10^8$ m s^{-1}/$(2.2 \times 10^{-18}$ J$)$
$= 9.04 \times 10^{-8}$ m

11 D Arrow length indicates size of energy change (ΔE)
Transitions shown consist of 3 of large energy and 2
of small energy
Since frequency $\propto \Delta E$, spectrum will contain 3 high
frequencies and 2 low frequencies

12 B $\lambda = h/p = h/mv$
$= 6.63 \times 10^{-34}$ J s$/(9.11 \times 10^{-31}$ kg $\times 3.0 \times 10^7$ m s$^{-1})$
$= 2.43 \times 10^{-11}$ m

13 A Doppler effect requires relative motion between
source and observer
If both moving with same velocity then there is no
relative movement between them

14 D Since apparent wavelength is larger (red-shift), galaxy
is moving away from Earth
$\Delta\lambda = (6.105 - 5.893) \times 10^{-7}$ m $= 2.12 \times 10^{-8}$ m
$v = c \times \Delta\lambda/\lambda = 3.0 \times 10^8$ m s$^{-1} \times 2.12 \times 10^{-8}$ m/
$(5.893 \times 10^{-7}$ m$) = 1.08 \times 10^7$ m s^{-1}

15 C

Practice questions

1 Incident radiation consists of packets of
energy/photons ✓
Photon energy > work function ✓
Radiation releases electrons which move to the anode,
forming a current ✓
All energy of a photon goes to one electron/one-to-one
interaction ✓
Potential difference opposes electron movement from
photocathode to anode ✓
Electrons have a range of kinetic energies so p.d. stops
the slowest first ✓
V_s stops all electrons reaching anode so no
current ✓ **Max 5**

(i) No effect on stopping potential ✓
(ii) 'a larger stopping potential would be needed ✓

2 Photon energy is too small/less than work function ✓

Threshold frequency $= 4.4 \times 10^{14}$ Hz ✓
work function $= hf_0 = 6.63 \times 10^{-34}$ J s $\times 4.4 \times 10^{14}$ Hz
$= 2.9 \times 10^{-19}$ J ✓
$= 2.9 \times 10^{-19}$ J$/(1.6 \times 10^{-19}$ J eV$^{-1}) = 1.8$ eV ✓

$hf = T + \phi$
$T = hf - \phi$ which is similar to $y = mx + c$ ✓
So graph of T against f should be a straight line ✓
With a negative T intercept equivalent to the work
function ✓ **Max 2**

Gradient of graph gives the Planck constant (in eV s) ✓

Line added to graph:
Starting from a higher frequency ($f_0 > 4.4 \times 10^{14}$ Hz) ✓
Parallel to original line (both have gradient h) ✓

3 Ionisation energy $= 10.4$ eV ✓
$= 10.4$ eV $\times 1.6 \times 10^{-19}$ J eV$^{-1} = 1.66 \times 10^{-18}$ J ✓

Downward 'sets' of arrows added to diagram:
From -1.6 eV directly to -10.4 eV ✓
From -1.6 eV first to -3.7 eV then to -10.4 eV ✓
From -1.6 eV first to -5.5 eV then to -10.4 eV ✓
From -1.6 eV first to -3.7 eV then to -5.5 eV
then to -10.4 eV ✓ **Max 3**

Photon energy $= hc/\lambda = 6.63 \times 10^{-34}$ J s $\times 3.00 \times 10^8$ m s^{-1}/
$(600 \times 10^{-9}$ m$)$ ✓
$= 3.31 \times 10^{-19}$ J$/(1.6 \times 10^{-19}$ J eV$^{-1})$ ✓
$= 2.07$ eV ✓
So need a downward transition between levels with an
energy difference of 2.07 eV
Transition is from -1.6 eV down to -3.7 eV ✓

4 $\frac{1}{2}mv^2 = 8.0 \times 10^{-21}$ J ✓
Mass of neutron $= 1.0087 \times 1.66 \times 10^{-27}$ kg
$v = \sqrt{[2 \times 8.0 \times 10^{-21}}$ J$/(1.0087 \times 1.66 \times 10^{-27}$ kg$)]$
$= 3090$ m s^{-1} ✓
Momentum $= mv = 1.0087 \times 10^{-27}$ kg $\times 3090$ m s^{-1} ✓
$= 5.2 \times 10^{-24}$ kg m s^{-1} ✓

$\lambda = h/mv = 6.63 \times 10^{-34} \text{ J s}/(5.2 \times 10^{-24} \text{ kg m s}^{-1})$
$= 1.3 \times 10^{-10} \text{ m}$ ✓

Yes ✓
As wavelength is similar to the spacing/size of
atoms/molecules ✓

5   Measure $\Delta\lambda$ for a known wavelength/spectral line ✓
Red-shift shows away and blue-shift shows towards/find
$v = c \times \Delta\lambda/\lambda$ ✓
Method only measures radial component of any
velocity ✓

$\Delta\lambda = v\lambda/c = 1.5 \times 10^7 \text{ m s}^{-1} \times 396.8 \text{ nm}/(3.0 \times 10^8 \text{ m s}^{-1})$ ✓
$= 19.8 \text{ nm}$ ✓
Apparent
wavelength = 396.8 nm + 19.8 nm = 416.6 nm ✓
$v = Hd$
$d = v/H = 1.5 \times 10^7 \text{ m s}^{-1}/(1.7 \times 10^{-18} \text{ s}^{-1})$ ✓
$= 8.8 \times 10^{24} \text{ m}$ ✓

t = distance
travelled/speed = $8.8 \times 10^{24} \text{ m}/(1.5 \times 10^7 \text{ m s}^{-1})$ ✓
$= 5.88 \times 10^{17} \text{ s}/(3.2 \times 10^7 \text{ s y}^{-1})$ ✓
$= 1.8 \times 10^{10} \text{ y}$ ✓
Speed/expansion rate may not have been constant ✓

Unit 5   Fields and forces and the A2 practical test

Part 1   Fields and forces

Quick test

1 D   From data, units of G are N m² kg⁻²
Since $F = ma$, N = kg m s⁻²
So N m² kg⁻² = kg m s⁻² m² kg⁻² = kg⁻¹ m³ s⁻²

2 B $g = GM/R^2$
$G = gR^2/M$

3 C   Initial centre-to-centre separation = $2r$
After moving $3r$, centre-to-centre separation = $5r$
Since $g \propto 1/r^2$, ratio is $5^2 : 2^2 = 25 : 4$

4 B $g = GM/R^2$
Since same density, mass $M \propto V \propto R^3$
So $g \propto R^3/R^2 \propto R$
$\frac{1}{2}R$ means $\frac{1}{2}g = \frac{1}{2} \times 16 \text{ N kg}^{-1} = 8 \text{ N kg}^{-1}$

5 C   Satellite A has a longer period so a higher orbit ($r^3 \propto T^2$)
Higher orbit so moving slower (see Worked Example 1)
So greater potential energy and smaller kinetic energy

6 D   The gravitational force is the only force acting on it
and keeps it in orbit

7 A $F = kq_1q_2/d^2$
So doubling both would quadruple F as would
halving d

8 C   Since electrostatic repulsion = gravitational attraction
$F = qq/4\pi\varepsilon_0 r^2 = Gmm/r^2$
$q^2/m^2 = 4\pi\varepsilon_0 G$
$q/m = \sqrt{(4\pi\varepsilon_0 G)}$

9 A   equipotentials are lines joining points with the same
potential energy

10 C $V = W/Q = 150 \text{ μJ}/(30 \text{ μC}) = 5 \text{ V}$

11 C $\frac{1}{2}mv^2 = eV \propto V$
So when V is ×4 and kinetic energy is ×4
v^2 is ×4 so v is ×2

12 B   With less resistance, initial current (V_s/R) will be
greater and discharge will be quicker

13 D   1 : 1/3 μF   2 : 3 μF   3 : 2/3 μF   4 : 1$\frac{1}{2}$ μF

14 A $W = \frac{1}{2}CV^2 = \frac{1}{2} \times 2 \times 10^{-6} \text{ F} \times (100 \text{ V})^2 = 0.01 \text{ J}$

15 B   Current-carrying wires have circular magnetic fields
around them
Currents into the page have clockwise fields

16 C   Since $F = BIl$, $B = F/Il$

17 D $B = \mu_0 nI = \mu_0 NI/L$
so $B \propto N$ and $B \propto 1/L$ and $B \propto I$

18 A   Both ends of solenoid will have same magnetic flux
density and $B = \mu_0 nI$ is unchanged

19 D   Opposition due to interaction of magnetic fields of
induced currents and permanent magnets
Slots reduced the possible current paths and so
currents are smaller

20 D $V_p/V_s = N_p/N_s$
$N_p = N_s V_p/V_s = 300 \times 230 \text{ V}/(69 \text{ V}) = 1000$

Practice questions

1   Force = $G \times$ mass$_1 \times$ mass$_2$/separation² ✓✓
$F = GMm/r^2$
$= 6.67 \times 10^{-11} \text{ N m}^2 \text{ kg}^{-2} \times 6.42 \times 10^{23} \text{ kg} \times 1 \text{ kg}/$
$(3.40 \times 10^6 \text{ m})^2$ ✓
$= 3.7 \text{ N}$ ✓

Smaller ✓
Since $R \propto g/\rho$
ρ for Mars = M/V
$= 6.42 \times 10^{23} \text{ kg}/[4\pi \times (3.40 \times 10^6 \text{ m})^3/3]$
$= 3900 \text{ kg m}^{-3}$ = similar to Earth's ✓
g for Mars (3.7 N kg⁻¹) is less than g for Earth ✓

2   Uniform gravitational field is one where the force acting
on a given mass is the same at all points ✓
Diagram (Figure A.15) showing:
Equally spaced vertical field lines ✓

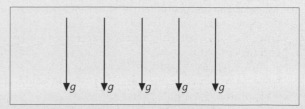

Fig A.15 *A uniform gravitational field*

With downward arrows ✓
Gravitational force on rocket decreases ✓
as rocket gets further from the centre of the Earth ✓

At D, gravitational force from Earth cancels that from Moon
Using $F = GMm/r^2$
$GM_{Earth}m/(346 \times 10^6 \text{ m})^2 = GM_{Moon}m/(38 \times 10^6 \text{ m})^2$
$M_{Earth}/M_{Moon} = (346 \times 10^6)^2/(38 \times 10^6)^2 = 83$

3 Nucleus has 92 protons, so $Q = 92 \times 1.6 \times 10^{-19}$ C ✓
$E = kQ/r^2 = 8.99 \times 10^9 \text{ N m}^2 \text{ C}^{-2} \times 92 \times 1.6 \times 10^{-19} \text{ C}/$
$\qquad (7.4 \times 10^{-15} \text{ m})^2$ ✓
$\qquad = 2.4 \times 10^{21} \text{ N C}^{-1}$ ✓

Electric field is away from the positive nucleus ✓

Similarity: both fields are radial/obey inverse square law/$\propto 1/r^2$ ✓
Difference: E field away, g field towards/E field attractive or repulsive, g field attractive only ✓

4 For a uniform electric field $E = V/d$
$V = Ed = 7.5 \times 10^5 \text{ V m}^{-1} \times 12 \times 10^{-2}$ m ✓
$\qquad = 9 \times 10^4$ V ✓

$\frac{1}{2}mv^2 = eV = 1.6 \times 10^{-19} \text{ C} \times 9 \times 10^4 \text{ J C}^{-1}$ ✓
$\qquad = 1.4 \times 10^{-14}$ J ✓

$v^2 = 2 \times 1.4 \times 10^{-14} \text{ J}/(9.11 \times 10^{-31} \text{ kg})$ ✓
$\qquad = 3.2 \times 10^{16} \text{ m}^2 \text{ s}^{-2}$
$v = \sqrt{(3.2 \times 10^{16} \text{ m}^2 \text{ s}^{-2})} = 1.8 \times 10^8 \text{ m s}^{-1}$ ✓

Since electron is negative, diagram similar to left-hand side of Figure 5.3 showing:
at least 3 equally spaced radial lines touching electron ✓
with arrows towards the electron ✓

5 Current is the rate of flow of charge or current = charge/time ✓

Since current is constant, charge on capacitor increases at a steady rate ✓
For a capacitor, $V \propto Q$ so V also increases at a steady rate, hence straight line ✓

When charged to 9.0 V,
$Q = CV = 4700 \times 10^{-6} \text{ F} \times 9.0 \text{ V} = 4.2 \times 10^{-2}$ C ✓
Since this takes 40 s, $I = Q/t = 4.2 \times 10^{-2} \text{ C}/(40 \text{ s})$ ✓
$\qquad\qquad\qquad = 1.06 \times 10^{-3}$ A ✓

Rheostat resistance must be decreased ✓
As capacitor charges, potential difference across rheostat decreases ✓
Since $I = V/R$, resistance must also decrease to keep current constant ✓

Initial current will be the same, but current will decrease with time
Capacitor will therefore take much longer to charge to 9.0 V
So add a second graph showing:
Curve from origin getting less steep but not peaking ✓
With same initial gradient as original straight line ✓

6 Slope = charge/potential difference = capacitance ✓

Area = $\frac{1}{2} \times$ charge $\times$ potential difference = energy stored ✓

Energy stored = $\frac{1}{2}CV^2 = \frac{1}{2} \times 100 \times 10^{-6} \text{ F} \times (250 \text{ V})^2$ ✓
$\qquad\qquad = 3.13$ J ✓

Power from cell = $IV = 0.20 \text{ A} \times 1.5 \text{ V} = 0.30$ W ✓
Time = energy/power = 3.125 J/(0.30 W) ✓
$\qquad\qquad = 10.4$ s ✓

7 Note that not all of Experiment 4 is needed here
General marking points are:
    system with a straight wire perpendicular to a magnetic field ✓
    method of providing and measuring direct current ✓
    method of measuring force ✓
    measure F for a range of values of I ✓
    plot F against I to get a straight line through the origin ✓

Upward magnetic force on wire = weight
$F = BIl = mg$ ✓
$I = mg/Bl = 1.5 \times 10^{-3} \text{ kg} \times 9.81 \text{ N kg}^{-1}/$
$\qquad (1.8 \times 10^{-5} \text{ T} \times 2.0 \text{ m})$ ✓
$\qquad = 409$ A ✓
Using Fleming's left-hand rule:
to produce an upward force with magnetic field to North, current must be from East to West ✓

409 A is a very large current and wire will get very hot/melt ✓

8 Vertical window is perpendicular to horizontal component of Earth's magnetic field ✓
Magnetic flux = BA = 20 µT $\times$ 1.3 m $\times$ 0.7 m ✓
$\qquad\qquad = 1.8 \times 10^{-5}$ T ✓

When opened through 90°, magnetic flux through window falls to zero
Induced e.m.f. = flux change/time = 1.8×10^{-5} T/(0.80 s) ✓
$\qquad\qquad = 2.3 \times 10^{-5}$ V ✓

9 Direction of any induced current is such as to oppose the change producing it ✓✓
Cylinder in T_3 takes longer since:
    magnetic flux cuts copper tube ✓
    e.m.f. induced ✓
    current in copper tube ✓
    produces magnetic field which opposes motion of magnet/slows magnet ✓
Since T_1 is plastic, no e.m.f./current induced ✓

Cylinder in T_2 is unmagnetised so no flux and no induction ✓
Both cylinders in T_1 and T_2 have only force of gravity acting on them ✓

Unit 6 Synoptic assessment

Part 1 Passage analysis

Practice questions

1 **(a)** Sketch showing Ganymede's magnetic field:
  shape similar to that of bar magnet ✓
  good quality, symmetric loops ✓
  spin axis marked at 10° to magnetic axis ✓
 (b) Sketch showing Ganymede's gravitational field
  (similar to Figure 5.1):
    radial shape ✓
    inward arrows ✓
    at least 8 field lines ✓
    symmetrical field lines ✓
  Measure positions at various times and calculate
  velocities ✓
  Calculate rate of change of velocity = $a = g$ ✓
 (c) (i) Unstable or decaying nuclei ✓
    that emit alpha, beta or gamma radiations ✓
  (ii) Induced e.m.f. or current ✓
    that opposes the change causing it ✓
  (iii) Motion of fluids i.e. convection currents ✓
    due to temperature differences and density
    changes ✓
  (iv) Interface between a solid and a
    liquid ✓ **Max 5**
 (d) Sketch graph showing:
    axes labelled with density and distance or
    radius ✓
    density decreasing with increasing distance ✓
    three distinct regions (iron core, silicate
    mantle, ice outer layer) ✓
  For the Earth, $I/mr^2 = 0.33$ ✓
  Mass of Earth $m = 4\pi r^3 \rho/3$ ✓
  $I = 0.33mr^2 = 0.33 \times 4\pi r^5 \rho/3$
    $= 0.44\pi \times (6400 \times 10^3 \text{ m})^5 \times 5500 \text{ kg m}^{-3}$
    $= 8.2 \times 10^{37} \text{ kg m}^2$ ✓
 (e) Ganymede 'formed some four billion years ago'
  which is 4 half-lives ✓
  so initial power = $2^4 \times$ present
  power = $2^4 \times 1.4 \times 10^{16}$ W ✓
      $= 16 \times 1.4 \times 10^{16}$ W $= 2.24 \times 10^{17}$ W ✓
  Tidal internal heating or friction at phase
  boundaries ✓
 (f) (i) $R = \rho l/A$
    l = circumference = $2\pi \times$ radius = $8\pi r$ and
    A = cross-sectional area = r^2 ✓
    $R = 1.1 \times 10^{-7} \, \Omega \text{ m} \times 8\pi \times 2.0 \times 10^5 \text{ m}/$
    $(2.0 \times 10^5 \text{ m})^2 = 1.38 \times 10^{-11} \, \Omega$ ✓
  (ii) e.m.f. = $IR = 3.0 \times 10^6 \text{ A} \times 1.38 \times 10^{-11} \, \Omega$ ✓
        $= 4.1 \times 10^{-5}$ V ✓

 (iii) Current flow produces a magnetic field ✓
  if the current were to decrease so would the
  magnetic field ✓
  so there would be a change in the magnetic
  flux through the ring ✓
  and an e.m.f. would be induced ✓
  which tries to keep the current
  flowing ✓ **Max 4**

2 **(a)** Any attempt to draw a relevant diagram ✓
  showing any **five** of the following correctly labelled:
    ocean or sea ✓
    evaporation ✓
    condensation or clouds ✓
    rain ✓
    reservoir or lake ✓
    HEP station or turbines ✓ **Max 5**
  Energy derived for the Sun or solar energy ✓
 (b) From gravitational potential energy ✓
  To mechanical or electrical energy ✓
  Tower diameter = 50 m so turbine diameter = 25 m
  and area $A = \pi r^2 = \pi \times (12.5 \text{ m})^2$ ✓
  Volume through turbine each second
  V' = speed $\times A = 20 \text{ m s}^{-1} \times 491 \text{ m}^2$ ✓
  Mass through turbine each second
  m' = density $\times V' = 650 \text{ kg m}^{-3} \times 9800 \text{ m}^3 \text{ s}^{-1}$ ✓
  GPE transferred each second = $m'g\Delta h$
    $= 6.4 \times 10^6 \text{ kg s}^{-1} \times 9.81 \text{ N kg}^{-1} \times 5000 \text{ m}$ ✓
    $= 3.13 \times 10^{11} \text{ J s}^{-1}$ or W ✓
 (c) T_1 is the temperature of the hot source and T_2 is the
  temperature of the cold sink ✓
  T_1 is the temperature of the sea ✓
  T_2 is the temperature of the upper atmosphere ✓
  Sea temperature = 10 °C (range allowed = 0 °C to
  30 °C) ✓
  $0.2 = (283 \text{ K} - T_2)/(283 \text{ K})$ ✓
  $T_2 = 283 \text{ K} - (0.2 \times 283 \text{ K})$
    $= 226 \text{ K}$ (range allowed = 218 K to 242 K) ✓
 (d) Weight or pull of Earth on MegaPower ✓
  upthrust from displaced sea water ✓
  tension in cables ✓
 (e) (i) Volume of cable = $\pi r^2 h$ ✓
    weight of cable = $\pi r^2 h \rho g$ ✓
    stress at top of
    cable = weight/area = $\pi r^2 h \rho g/\pi r^2 = h\rho g$ ✓
    so h = stress/ρg and cable will break if
    $h >$ maximum stress/ρg ✓ **Max 3**
  (ii) Unit of stress σ is N m^{-2} ✓
    unit of density ρ is kg m^{-3} and unit of g is
    N kg^{-1} ✓
    so unit of $\sigma/\rho g$ is
    N m^{-2}/(kg m^{-3} × N kg^{-1}) = N m^{-2}/
    (N m^{-3}) = m ✓
 (f) No toxic waste gases ✓
  No fuel used or it is a renewable resource ✓
  Any two possible problems from:
    leaks of ammonia
    aircraft or shipping hazard
    it will lower the sea temperature
    cables breaking in a storm ✓✓ **Max 3**

Part 2 Synthesis material

Quick test

1 A $W = \frac{1}{2}Fx = \frac{1}{2} \times 15$ N $\times 25 \times 10^{-3}$ m $= 0.19$ J
$W = \frac{1}{2}VQ = \frac{1}{2} \times 9$ V $\times 1 \times 10^{-3}$ C $= 0.0045$ J
$W = \frac{1}{2}kx^2 = \frac{1}{2} \times 20$ N m^{-1} $\times (6 \times 10^{-2}$ m$)^2 = 0.036$ J
$W = \frac{1}{2}CV^2 = \frac{1}{2} \times 500 \times 10^{-6}$ F $\times (5$ V$)^2 = 0.0063$ J

2 C The direction of the electric field is **away from** the positive charge

3 D For small charged particles electrostatic forces dominate (hence 2 protons repel rather than attract) The smaller the distance apart, the larger the force

4 B $C = $ time constant$/R$
$= 1.0 \times 10^{-3}$ s$/(1.0 \times 10^3 \ \Omega) = 1.0 \times 10^{-6}$ F $= 1$ μF
$V = Q/C = 10$ μC$/(1$ μF$) = 10$ V
After 2.0 ms, $Q = (1/e) \times (1/e) \times 10$ μC $= 1.4$ μC
Initial
current $= V/R = 10$ V$/(1 \times 10^3 \ \Omega) = 1 \times 10^{-2}$ A $= 10$ mA

5 B Time constant of decay $= 1/\lambda = t_{\frac{1}{2}}/\ln 2 = 5 \times 10^7$ s$/\ln 2$

6 D 14.011 179 u > 14.003 074 u and excess mass emitted as energy

7 C $^{32}_{16}$S $+ ^{94}_{42}$Mo $= ^{126}_{58}$X $= ^{122}_{56}$Ba $+ ^{4}_{2}$He

8 A Charge of alpha
particle $= 2 \times 1.6 \times 10^{-19}$ C $= 3.2 \times 10^{-19}$ C
$F = BQv = 2.0$ T $\times 3.2 \times 10^{-19}$ C $\times 1.6 \times 10^7$ m s^{-1}
$= 1.0 \times 10^{-11}$ N

9 D $BQv = mv^2/r$ so $r = mv/BQ$
new $r = m \times 2v/(\frac{1}{2}B \times Q) = 4 \times$ old r

10 C New $r = m \times 2v/BQ = 2 \times$ old r so second particle has to travel twice as far
Since travelling at twice the speed it takes the same time $t = 2x/2v$

11 A

12 B If Q decreased then inward force would be less than that required so particle would spiral outwards
If v decreased then inward force would be more than that required so particle would spiral inwards
If B decreased then inward force would be less than that required so particle would spiral outwards
If m increased then inward force would no longer be sufficient so particle would spiral outwards

Practice questions

1 $(1/e) \times 9.0$ V $= 3.3$ V ✓
Time taken to fall to 3.3 V $= 23$ s ✓
Or time taken to fall to 4.5 V (half-life) $= 16$ s
time constant $= 16$ s$/\ln 2 = 23$ s

$R = V/I = 9.0$ V$/(0.19 \times 10^{-3}$ A$)$ ✓
$= 4.7 \times 10^4 \ \Omega$ ✓

$C = $ time constant$/R = 23$ s$/(4.7 \times 10^4 \ \Omega)$ ✓
$= 4.9 \times 10^{-4}$ F ✓

Line added to graph:
curve starting from origin and getting less steep ✓
and reaching roughly 7.5 V (i.e. 9.0 V $-$ 1.5 V) at
$t = 40$ s ✓

2 Total mass of particles
before $= 1.008$ 70 u $+ 235.043$ 94 u $= 236.052$ 64 u
Total mass of particles
after $= 143.922$ 85 u $+ 91.926$ 27 u $= 235.849$ 12 u
Mass of energy
released $= 236.052$ 64 u $- 235.849$ 12 u $= 0.203$ 52 u ✓
$= 0.203$ 52 $\times 1.66 \times 10^{-27}$ kg $= 3.38 \times 10^{-28}$ kg ✓
Energy
released $= c^2 \Delta m = (3 \times 10^8$ m s$^{-1})^2 \times 3.38 \times 10^{-28}$ kg ✓
$= 3.04 \times 10^{-11}$ J ✓

3 Paths added to Figure 6.12 showing:
beam of protons from left curving upwards in field ✓

beam of protons from right curving downwards in field ✓
with a smaller radius (since slower) ✓

Magnetic field drawn perpendicular to current and to cross-section (Figure A.16) ✓✓

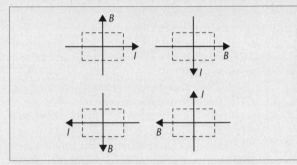

Fig A.16 *Each of these arrangements produces a force out of the page*

Correct orientation to give force out of page ✓

Part 3 Synoptic questions

Practice questions

1 (a) From kinetic energy ✓
to elastic or strain energy ✓

(b) (i) $p_1V_1 = p_2V_2$ ✓
$p_2 = p_1V_1/V_2 = 116$ kPa $\times V/0.84V = 138$ kPa ✓
$\Delta p = 138$ kPa $- 116$ kPa $(= 22$ kPa$)$ ✓

(ii) Pressure increases since gas molecules are closer together ✓
So have more collisions with the walls of the container ✓
Speed/kinetic energy of molecules increases with temperature ✓
So molecules hit the walls harder/undergo a greater change of momentum ✓
Max 6 for (b)

(c) (i) Use $c = \sqrt{(2\pi a^3 \Delta p/m)}$
With $a = 33.5 \times 10^{-3}$ m and
$m = 57.5 \times 10^{-3}$ kg ✓
and $\Delta p = 32 \times 10^3$ Pa (range allowed $= 22$ kPa to 38 kPa) ✓

So
$c = \sqrt{[2\pi \times (33.5 \times 10^{-3} \text{ m})^3 \times 32 \times 10^3 \text{ Pa}/(57.5 \times 10^{-3} \text{ kg})]}$
$= 11.5 \text{ m s}^{-1}$ (range allowed $= 9.5 \text{ m s}^{-1}$ to 12.5 m s^{-1}) ✓

Pulse travels full
circumference $= 2\pi r = 2 \times \pi \times 33.5 \times 10^{-3}$ m
$= 0.210$ m ✓

Time of contact
$t_c = x/v = 0.210 \text{ m}/(11.5 \text{ m s}^{-1}) = 0.018$ s ✓

(ii) Measurement of time of contact:
Device used: e.g. pressure pad on wall or narrow beam of light along wall ✓
How device works: e.g. conducts when compressed or light gate reacts to broken beam ✓
How time found: e.g. connected to scalar timer ✓

2 (a) (i) and **(iii)** field lines and equipotentials added (see Figure A.17) with:

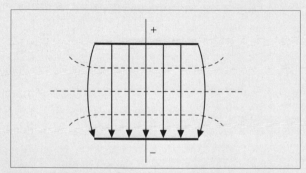

Fig A.17 *Field lines and equipotentials are perpendicular*

central field line with arrow down, from $+$ to $-$ ✓
other field lines on either side ✓
edge field lines showing 'bulges' ✓

central (3.0 V) equipotential horizontal ✓
two more equipotentials drawn with correct curvature for field lines drawn ✓

(ii) $E = V/d = 6.0 \text{ V}/(15 \times 10^{-2} \text{ m})$ ✓
$= 40 \text{ V m}^{-1}$ or N C^{-1} ✓

(b) (i) Potential at X $= 3.0$ V (since half-way between two electrodes) ✓
Potential at Y $= 3.0$ V (since resistors are equal) ✓
so there is no potential difference across milliammeter ✓

(ii) e.g. replace milliammeter with voltmeter connected to 0 V instead of Y ✓
move probe over the paper and record potentials ✓
or replace milliammeter with voltmeter add 3.0 V to all voltmeter readings
or change resistor ratio and calculate potential at Y
find points on paper where milliammeter reads zero

(c) (i) $R = \rho l/A$ ✓
Where $l = x$ and $A = xt$ ✓
so $R = \rho x/xt = \rho t$

(ii) $\rho = Rt$ ✓
$= 1000 \ \Omega \times 0.14 \times 10^{-3} \text{ m} = 0.14 \ \Omega \text{ m}$ ✓

3 (a) Unit of $gd = \text{m s}^{-2} \times \text{m} = \text{m}^2 \text{ s}^{-2}$ ✓
so unit of $\lambda = \text{s} \times \sqrt{(\text{m}^2 \text{ s}^{-2})} = \text{s} \times \text{m s}^{-1} = \text{m}$ ✓

$c = f\lambda = \lambda/T$ ✓
$c = \lambda T = \sqrt{(gd)}$ ✓
$= \sqrt{(9.81 \text{ m s}^{-2} \times 2200 \text{ m})} = 147 \text{ m s}^{-1}$ ✓
or $\lambda = T\sqrt{(gd)} = (18 \times 60) \text{ s} \times \sqrt{(9.81 \text{ m s}^{-2} \times 2200 \text{ m})}$
$= 1.59 \times 10^5$ m
$c = x/t = \lambda/T$
$= 1.59 \times 10^5 \text{ m}/[(18 \times 60) \text{ s}] = 147 \text{ m s}^{-1}$

(b) Description of ripple tank experiment including:
use of two sources ✓
which are coherent e.g. 2 dippers, 2 slits, source and reflector ✓
producing superposition/maximum and minimum pattern ✓
how pattern observed e.g. lamp to project, use of strobe ✓
measure path difference to n^{th} maximum ✓
and divide by n to get λ ✓
or measure across several sets of nodes/antinodes
and use node-node $= \frac{1}{2}\lambda$

(c) (i) Energy lost by conduction/convection/evaporation ✓
Energy spread over large/varying area ✓
or water has a very high specific heat capacity

(ii) Tides vary position of energy extraction/amount of energy varies with wave size ✓
Energy would be converted to electrical energy which is hard to store ✓
Max 3 for (i) & (ii)

(iii) Electrical power $= 1.6\% \times 2.0 \text{ MW} = 32 \text{ kW}$ ✓
In 24 h, energy $= 32 \text{ kW} \times 24 \text{ h} = 770 \text{ kW h}$ ✓

4 (a) (i) Energy per electron $= eV$ so energy per second $= NeV$ ✓

(ii) Energy supplied $= Pt = 2.4 \text{ W} \times 20 \text{ s}$ ✓
$= 48$ J ✓
Energy transfer $= mc\Delta\theta$
$m = 48 \text{ J}/(730 \text{ J kg}^{-1} \text{ K}^{-1} \times 85 \text{ K})$ ✓
$= 7.7 \times 10^{-4}$ kg ✓
Assumptions:
All energy transferred to heat/none transferred to light ✓
No heat conducted away from spot/only spot heated ✓

(b) Sketch/description of apparatus (similar to Figure 2.44) ✓
Measure IVt ✓
Measure m and $\Delta\theta$ ✓
Difficulties:
Glass is a poor conductor so only that near heater gets heated ✓

Heat as likely to enter surroundings as rest of glass ✓

(c) (i) One cycle = 2.5 divisions

= 2.5 × 100 μs = 2.5 × 10^{-4} s ✓

$f = 1/T = 1/(2.5 \times 10^{-4} \text{ s}) = 4000$ Hz ✓

(ii) Signal rises 1.6 divisions each cycle = 1.6 × 0.2 V = 0.32 V ✓

Rate of rise = 0.32 V/(2.5 × 10^{-4} s)

= 1280 V s^{-1} ✓

Appendix 1 General requirements

 ## Introduction

The general requirements consist of material common to the whole of physics. The part of the specification containing it is easily overlooked, as it comes before the start of Unit 1. However, it is very important as most of its content may be tested in any of the assessment tests, although a small amount only applies to tests PHY4, 5/01, 5/02 and 6.

 ## Things to learn and understand for all assessment tests

Physical quantities

- A physical quantity is any measurable physical property and is written as the product of a number and a unit

- All measurements must be made using an internationally agreed unit to allow comparisons between measurements from different sources ... the International System of Units (SI)

- The prefixes that are used with units to show larger or smaller multiples:

Prefix	Symbol	Multiplier
giga	G	10^9
mega	M	10^6
kilo	k	10^3
milli	m	10^{-3}
micro	μ	10^{-6}
nano	n	10^{-9}
pico	p	10^{-12}

- The approximate sizes of common physical quantities e.g. mass of an adult ~ 70 kg and diameter of an atom ~10^{-10} m

- There has to be a starting set of base physical quantities, together with a corresponding set of base units

- Length, mass, time, current, temperature interval and amount of substance are all base SI physical quantities

- All other physical quantities can be produced by combining base physical quantities and are called derived physical quantities

- The metre, the kilogram, the second, the ampere, the kelvin and the mole are all base units
- All other units can be produced by combining base units and are called derived units
- A word equation is simply an equation written using words instead of symbols that can be used to define either a derived physical quantity or its derived unit e.g. *charge = current × time* and *coulomb = ampere × second*
- All equations must be homogeneous in that they must equate, add together or subtract the same type of physical quantity
- Homogeneous equations can still be wrong as they may contain incorrect numbers

Scalars and vectors

- All physical quantities are either scalars or vectors, where a scalar quantity has only size whereas a vector quantity has both size and direction
- A single vector can be split up into any number of components although it is usual to choose two components that are at right angles to each other, a process called resolution of the vector
- Any number of vectors can be added together to find the single resultant vector that could be used to replace them by the use of either a scale drawing or a suitable calculation

Graphs

- The plotting of clear and accurate graphs is an essential skill in physics; make sure you practise selecting suitable, sensible scales and plotting accurate data points ... you will definitely have to plot graphs in tests PHY3/02, 5/02 and 6
- A graph can replace a data table; you should be able to extract data from either of these formats
- A straight line graph represents a linear relationship and has an equation of the form $y = mx + c$ where m is the slope and c is the intercept
- The slope of a graph gives the rate at which one quantity is changing with the other; for a straight line this will be the same all the way along it, while it varies from place to place along a curve
- Drawing a tangent to a curve and finding its slope gives the instantaneous rate of change at that point
- The area enclosed by the line and the *x*-axis *may* represent a physical quantity, e.g. work done = area enclosed by a force–displacement graph
- Sketch graphs are an excellent way of showing how one physical quantity varies with another; a single sketch graph can replace many words and often result in a better answer

 Checklist for all tests

Before attempting any of the questions in Units PHY1 to PHY6, check that you:

- ☐ know that a physical quantity is any measurable physical property and is written as the product of a number and an internationally agreed unit

❑ know the unit of every physical quantity that you encounter and remember to add the correct unit to all your calculated answers

❑ have learnt the prefixes used with units to show larger or smaller multiples

❑ understand the system of base and derived physical quantities used in physics

❑ know the six base SI physical quantities included in the specification and their associated base units

❑ can use word equations to define derived physical quantities and their derived units

❑ know how to express a derived unit as a combination of only base units

❑ know how to use homogeneity to check the possible correctness of an equation while appreciating why a homogeneous equation may still be wrong

❑ can explain the difference between a scalar and a vector quantity

❑ know how to resolve a vector into two components that are at right angles to each other

❑ can find the resultant of several vectors using a scale drawing and of two perpendicular vectors by calculation

❑ have practised selecting suitable scales and plotting data points

❑ can read numerical data from a graph

❑ know the equation $y = mx + c$ and its relationship to a straight line graph

❑ can find the slope m and the intercept c of a straight line graph and appreciate that both these quantities are likely to have units

❑ can find an accurate value for the instantaneous rate of change of one quantity with another in situations where they vary both linearly and non-linearly

❑ are familiar with the concept of rate of change with time, as found in the definitions of velocity, acceleration, force, activity etc.

❑ can find the area enclosed by the line and the x-axis by an appropriate method e.g. by direct calculation for a straight line and by counting squares for a curve

❑ will remember to use a sketch graph whenever it helps me produce a better answer

Additional things to learn and understand for tests PHY4, 5/01, 5/02 and 6

Graphs

● Know the shapes of graphs of simple functions (Figure A1) so that you can quickly sketch the different relationships between the physical quantities that you meet with in your studies

> Graph A does not touch either axis ... it might represent how the pressure of a fixed mass of gas varies with its volume at constant temperature
> Graph B rises from the origin with an increasing slope ... it might represent how the displacement of a uniformly accelerating body varies with time

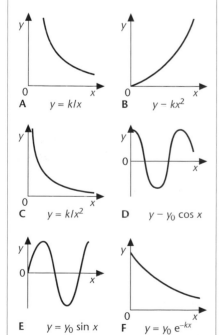

A $y = k/x$ B $y = kx^2$

C $y = k/x^2$ D $y = y_0 \cos x$

E $y = y_0 \sin x$ F $y = y_0 e^{-kx}$

Fig A1 *Graph shapes to learn*

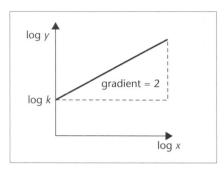

Fig A2 *log y = 2 log x + log k so*
y = kx²

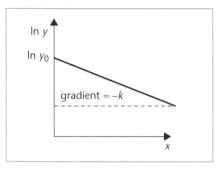

Fig A3 *ln y = –kx + ln y₀ so y = y₀ e⁻ᵏˣ*

Graph C is similar to graph A but is less shallow ... it might represent how the intensity varies with distance from a point source of light

Graph D is a co-sinusoidal curve about a zero position ... it might represent how the displacement of a heavy oscillating pendulum varies with time

Graph E is a sinusoidal curve about a zero position ... it might represent how an alternating voltage varies with time

Graph F is an exponential decay curve; it falls from a positive intercept with a decreasing slope, never quite reaching the x-axis ... it might represent how the voltage across a discharging capacitor varies with time

● Know how to use your calculator to find logarithms

● Know that a logarithmic plot can be used to find the power of a relationship, e.g. a logarithmic plot of the data of graph A in Figure A1 would produce a linear graph with a slope of –1 (since $k/x = kx^{-1}$ i.e. x is raised to a power of –1), whereas a slope of +2 (Figure A2) would be produced from the data of graph B (kx^2 i.e. x is raised to a power of 2)

● Know that if the natural logarithm of the data represented by the y-axis in graph F is plotted against the x-axis data, a straight line graph is produced proving that the original curve is exponential (Figure A3)

Appendix 2 List of data, formulae and relationships

Appropriate parts of this list will be printed at the back of each of your unit tests.

Data

Speed of light in vacuum	$c = 3.00 \times 10^8$ m s^{-1}
Gravitational constant	$G = 6.67 \times 10^{-11}$ N m^2 kg^{-2}
Acceleration of free fall	$g = 9.81$ m s^{-2}
Gravitational field strength	$g = 9.81$ N kg^{-1}
Electronic charge	$e = -1.6 \times 10^{-19}$ C
Electronic mass	$m_e = 9.11 \times 10^{-31}$ kg
Electronvolt	$1 \text{ eV} = 1.6 \times 10^{-19}$ J
Planck constant	$h = 6.63 \times 10^{-34}$ J s
Unified atomic mass unit	$u = 1.66 \times 10^{-27}$ kg
Molar gas constant	$R = 8.31$ J K^{-1} mol^{-1}
Permittivity of free space	$\varepsilon_0 = 8.85 \times 10^{-12}$ F m^{-1}
Coulomb law constant	$k = 1/4\pi\varepsilon_0 = 8.99 \times 10^9$ N m^2 C^{-2}
Permeability of free space	$\mu_0 = 4\pi \times 10^{-7}$ N A^{-2}
Stefan–Boltzmann constant	$\sigma = 5.67 \times 10^{-8}$ W m^{-2} K^{-4}

Rectilinear motion
For uniformly accelerated motion:

$$v = u + at$$
$$x = ut + \tfrac{1}{2}at^2$$
$$v^2 = u^2 + 2ax$$

Forces and moments

Moment of F about O = $F \times$ (perpendicular distance from F to O)
Sum of clockwise moments about any point in a plane = Sum of anticlockwise moments about that point

Dynamics

Force	$F = m\Delta v/\Delta t = \Delta p/\Delta t$
Impulse	$F\Delta t = \Delta p$

Mechanical energy

Power	$P = Fv$

Radioactive decay and the nuclear atom
Activity $\qquad$ $A = \lambda N$ (decay constant λ)
Half-life $\qquad$ $\lambda t_{\frac{1}{2}} = 0.69$

Electrical current and potential difference
Electric current $\qquad$ $I = nAQv$
Electric power $\qquad$ $P = I^2 R$

Electrical circuits
Terminal potential difference $\quad V = \varepsilon - Ir$ (e.m.f. ε; internal resistance r)
Circuit e.m.f. = ΣIR
Resistors in series $\qquad$ $R = R_1 + R_2 + R_3$
Resistors in parallel $\qquad$ $1/R = 1/R_1 + 1/R_2 + 1/R_3$

Heating matter
Change of state: $\qquad$ energy transfer = $L\Delta m$ (specific latent heat L)
Heating and cooling: $\qquad$ energy transfer = $mc\Delta T$ (specific heat capacity c,
$\qquad$ temperature change ΔT)
Celsius temperature $\qquad$ $\theta/°C = T/K - 273$

Kinetic theory of matter

$\qquad$ $T \propto$ average kinetic energy of molecules

Kinetic theory $\qquad$ $p = \frac{1}{3}\rho <c^2>$

Conservation of energy
Change of internal energy $\quad \Delta U = \Delta Q + \Delta W$ (energy transferred thermally
$\qquad$ ΔQ, work done on body ΔW)
Efficiency of energy transfer = useful output/input
For a heat engine, maximum efficiency = $(T_1 - T_2)/T_1$

Astrophysics

$\qquad$ $L = \sigma T^4 \times$ surface area
$\qquad$ $\lambda_{max} T = 2.898 \times 10^{-3}$ m K
$\qquad$ surface area of a sphere = $4\pi r^2$
$\qquad$ intensity = $L/4\pi D^2$

Solid materials

$\qquad$ $F = k\Delta x$
$\qquad$ $\sigma = F/A$
$\qquad$ $\varepsilon = \Delta l/l$
$\qquad$ $E =$ stress/strain

Work done in stretching $\Delta W = \frac{1}{2}F\Delta x$ (provided Hooke's law holds)
Energy density = energy/volume

Nuclear and particle physics

$\qquad$ $r = r_0 A^{1/3}$

Medical physics
Effective half-life $\qquad$ $1/t_e = 1/t_r + 1/t_b$
Acoustic impedance $\qquad$ $Z = c\rho$
Reflection coefficient = $(Z_1 - Z_2)^2/(Z_1 + Z_2)^2$

Circular motion and oscillations
Angular speed $\omega = \Delta\theta/\Delta t = v/r$ (radius of circular path r)
Centripetal acceleration $a = v^2/r$
Period $T = 1/f = 2\pi/\omega$ (frequency f)
Simple harmonic motion:

> displacement $x = x_0 \cos 2\pi ft$
> maximum speed $= 2\pi fx_0$
> acceleration $a = -(2\pi f)^2 x$

For a simple pendulum $T = 2\pi\sqrt{(l/g)}$
For a mass on a spring $T = 2\pi\sqrt{(m/k)}$

Waves
At distance r from a point source of power P intensity $I = P/4\pi r^2$

Superposition of waves
For interference of light using two slits of slit separation s

> wavelength $\lambda = xs/D$ (fringe width x, slits to screen distance D)

Quantum phenomena
Photon model $E = hf$
Maximum energy of photoelectrons $= hf - \phi$ (work function ϕ)
Energy levels $hf = E_1 - E_2$
de Broglie wavelength $\lambda = h/p$

Observing the universe

> $\Delta f/f = \Delta\lambda/\lambda = v/c$ OR $2v/c$
> $v = Hd$

Gravitational fields
Gravitational field strength $g = F/m$
for radial field $g = Gm/r^2$ (numerically)

Electric fields
Electric field strength $E = F/Q$
for radial field $E = kQ/r^2$ (where for free space or
                                                          air $k = 1/4\pi\varepsilon_0$)
for uniform field $E = V/d$
For an electron in a vacuum tube $e\Delta V = \Delta(\frac{1}{2}m_e v^2)$

Capacitance
Energy stored $W = \frac{1}{2}CV^2$
Capacitors in parallel $C = C_1 + C_2 + C_3$
Capacitors in series $1/C = 1/C_1 + 1/C_2 + 1/C_3$
Time constant for capacitor to charge or discharge $- RC$

Magnetic fields
Force on a wire $F = BIl$

> $B = \mu_0 nI$
> $B = \mu_0 I/2\pi r$

Magnetic flux $\Phi = BA$
E.m.f. induced in a coil $\varepsilon = -N\Delta\Phi/\Delta t$ (number of turns N)

Accelerators

Mass-energy	$\Delta E = c^2 \Delta m$
Force on a moving charge	$F = BQv$

Analogies in physics

$$Q = Q_0\, e^{-t/RC}$$
$$N = N_0\, e^{-\lambda t}$$

Activity	$A = \lambda N$
Exponential decay	$t_{\frac{1}{2}}/RC = \ln 2$

$$t_{\frac{1}{2}} = \ln 2/\lambda$$

Experimental physics

Percentage uncertainty = estimated uncertainty $\times$ 100%/average value

Mathematics

$$\sin (90° - \theta) = \cos \theta$$
$$\ln (x^n) = n \ln x$$
$$\ln (e^{kx}) = kx$$

Equation of a straight line	$y = mx + c$
Surface area	cylinder $= 2\pi rh + 2\pi r^2$
	sphere $= 4\pi r^2$
Volume	cylinder $= \pi r^2 h$
	sphere $= 4\pi r^3/3$
For small angles:	$\sin \theta \approx \tan \theta \approx \theta$ (in radians)
	$\cos \theta \approx 1$

Index